"Philosophy is written in this grand book,
the universe, which stands continually
open to our gaze."

GALILEO GALILEI

WILDSAM FIELD GUIDES™

Copyright © 2019

Published in the United States
by Wildsam Field Guides, Austin, Texas.

ISBN 978-0-578-59418-7

Design by Alan Kahler
Illustrations by Jamison Harper

To find more field guides, please visit
www.wildsam.com

# CONTENTS

*A history of man and the moon, through science,
exploration, history, spirituality, geopolitics and prose*

# WELCOME

**MY WIFE AND I HAVE TWO SONS.** Booker is four years old; Truett will be two next spring. This morning, as we hurried the day, Booker sat quietly at our kitchen table drawing the solar system. The sun glowed on the black paper. Mercury was a wild crayon orange, too. Venus a pinkish-purple. Earth had its green hills and blue waves, and the moon was an equally-sized, silvery swirl. Booker was adding a lunar landing module [his words], when I looked over his shoulder, enough to hear his soft slow breathing. *Look dada*, he whispered. *The moon.*

If you ask me—*Why did Wildsam publish a book about the moon?*—my easiest answer is right here, sitting atop the kitchen table. He's holding my hand as we walk around the Saturn V. He's up in that rickety observatory tower with me, looking at the rising harvest glow. He's in my arms, well after dark, and he's whispering the same words: *Look dada, the moon.*

There's an intimacy with the moon, a closeness inherent. "Look at her," Borges writes. "She is your mirror." Or Shakespeare: "Like a silver bow new-bent in heaven, shall behold the night of our solemnities." And Emily Dickinson, who brings her to life: "She turns her perfect face upon the world below." Unlike the sun, the moon invites us to behold. The French philosopher Simone Weil affirms: "It represents the beauty of God."

A few years ago, much of America found the path of a total eclipse. Beyond the celestial miracle that allows for this moment—the sun four hundred times farther from Earth than the moon, the moon four hundred times smaller than the sun, an impossible bit of chance—what stuck with me that day was how instinctive it felt to gather. Midtown street, Tennessee field, highway rest stop.

Our world is a complicated, often brutal little orb. As I've rediscovered the moon this year, I've drawn close to a simple truth: There is one moon. Its borrowed light covers us all. When I talk to the moon, you out there, across the ocean, might be whispering up the same words. Draw near. —*Taylor Bruce*

ESSENTIALS

A short primer on lunar science and selenography, historic
moments and notable entry points of interaction

# ORIGINS

*Imagine a rock the size of Mars. It has a name: Theia, mother of the moon goddess in Greek mythology. The rock hurtles through space until, one day, about four and a half billion years ago, it slams into proto-Earth. Violence follows, the molten debris ring concentrates in orbit around the planet.*

**THE MOON IS BORN.**

**AGE**
4.6 billion years

**DIAMETER**
2,158 miles across
*Roughly New York City to
Grand Canyon NP*

**MASS**
81.5 Moons would equal
Earth's total mass

**DISTANCE FROM EARTH**
252,716 miles [*Apogee*]
221,468 miles [*Perigee*]

**ORBITAL GROWTH**
3.8 centimeters per year
*In 50 million years, the
Moon's orbit will double*

**SURFACE TEMP (DAY)**
273° F [134° C]

**SURFACE TEMP (NIGHT)**
-254° F [-154° C]

**ORBITAL SPEED**
22,887 mph

**GRAVITY**
16.6% of Earth
*150 lbs becomes 25 lbs*

**LENGTH OF LUNAR DAY**
27.3 Earth days

**DARKNESS**
Near the lunar poles, there are
regions in permanent shadow
with temps colder than Pluto

# ATMOSPHERE

Comparable in density to Earth's outermost atmosphere, the moon is unprotected from cosmic rays, solar winds and meteorites. No soundwaves exist in the nonexistent air. A near constant cloud of moon dust floats, caused by five tons of comet particles striking every day.

# PHASES

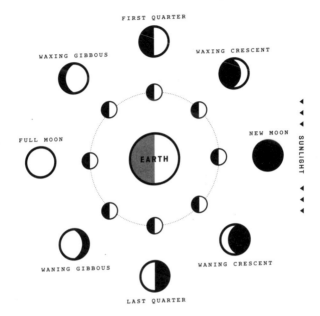

FIRST QUARTER

WAXING GIBBOUS

WAXING CRESCENT

FULL MOON

EARTH

NEW MOON

SUNLIGHT

WANING GIBBOUS

WANING CRESCENT

LAST QUARTER

### THE DARK SIDE

*Is not, in fact, dark—it's just the side we never
see. The moon rotates on its axis at close to the same
speed as its orbit of Earth, so the near side, or earth-
ward side is constantly facing us. The far side of the
moon receives just as much sunlight as the near side,
we just never get to glimpse it from where we are.*

# GEOLOGY

*The moon is a rugged and vacant expanse of dead volcanoes, miles-wide craters, lava flows and mountainous ridges. It is magnificent desolation, as Buzz Aldrin famously observed. This severe terrain formed billions of years ago through molten eruptions and a period of heavy bombardment experienced throughout the solar system. But because it lacks a system of plate tectonics and atmospheric-driven erosion, the moon's surface preserves the history of our cosmos, visible even with the naked eye.*

......................................................................................

| | |
|---|---|
| Highlands | From Earth, the lighter parts of the moon are the highland regions, called "terra" by early astronomers. The highlands contain greater density of large craters than the low-lying maria. The moon's highest point is not named, and not a mountain, but rather a gradually sloping region quite possibly formed by debris from the impact crater in the Aitken basin. It tops out at 35,387 feet. |
| Maria | Dried pools of basaltic lava make up these wide lunar "seas," once believed to be filled with water due to their dark appearance. Maria cover roughly one-third of the moon's near side. 17th-century Italian astronomer-priest Giovanni Riccioli named these larger areas for weather, hence the Sea of Tranquility and Ocean of Storms. |
| Impact Craters | Formed by asteroids or comets slamming into the moon, craters vary in size and shape, from microscopic divots to rimmed depressions twice the size of Connecticut. Most craters are named for explorers and scientists. There are more than 300,000 craters [no smaller than half a mile wide] on the moon's near side. |

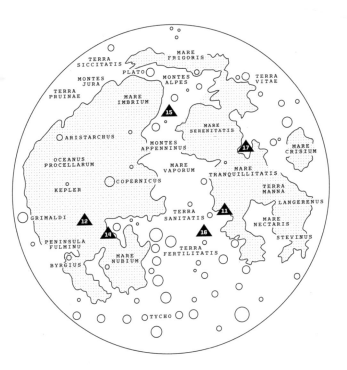

*Only six manned missions have successfully landed
on the moon, each one part of NASA's Project Apollo.
The last, Apollo 17, landed December 11, 1972.*

# HUMANS AND THE MOON

15,000 BC .... Lascaux cave paintings in Dordogne region of France depict seasons and 13 moons of lunar year

8,000 BC ...... First known lunar calendar created with series of pits in Scotland

300 BC ......... Babylonian temple priests develop a model, called System B, to predict moon's movement

190 BC ......... Theory of the moon and tides put forth by Greek astronomer Seleucus

37 BC ........... Chinese astronomer posits that the moon's brightness is due to reflective light from Sun

150 ............. Ptolemy's geocentric treatise *The Almagest* allows for estimates of moon size, distance

632 ............. Prophet Muhammad's splitting of the moon miracle recorded in the Quran

1364 ........... First modern astrarium clock by Giovanni de' Dondi, tracks the sun, moon and planets

1543 ........... Copernicus proposes heliocentric model of universe, putting moon in orbit of Earth

1610 ........... With telescopic aid, Galileo publishes detailed sketches of mountainous moon

1651 ........... Jesuit priest Giovanni Riccioli publishes a detailed lunar map, establishing the naming system in use today

1687 ........... Newton's laws of motion, theory of gravity provide near-perfect model of moon's motion

1829 ........... German astronomer, Franz von Paula Gruithuisen, suggests that lunar craters result of asteroid impacts

1898 ........... George Darwin, son of Charles, suggests Earth and the moon were once single body

1959 ........... Soviet Luna 3 takes first photos of the dark side of the moon

1967 ........... Outer Space Treaty signed by 109 nations ensures no weapons on the moon

1969 ........... Apollo 11 brings back 50 moon rock samples, confirming no living organisms

1981 ........... MTV launches its channel with footage of Apollo 11; Moonman becomes its branded trophy

2019 ........... China lands lunar probe on the dark side of the moon

# GLOSSARY

### ARMALCOLITE
Mineral discovered by Apollo 11, coined using letters of astronauts names

### EARTHSHINE
Sunlight reflected by the Earth that illuminates the darkest parts of the moon

### IMPACT CRATERS
Over 300,000 depressions formed by asteroid impact on near side alone

### INTERCALATION
Insertion of days [or months] into calendar to create rhythm between the moon and the seasons

### METONIC CYCLE
The 19 years—or 6,940 days—it takes for the moon to reappear in exactly the same part of the sky

### SYNODIC MONTH
The exact period between successive new moons: 29.531 days

### SYZYGY
Three or more celestial bodies in near-straight line, from Greek "yoked together"

### TRANSIENT LUNAR PHENOMENA
Sudden changes in color or luminosity of the moon, perhaps from gas release or impact events

# MOON TYPES

### SUPERMOON
New moon at perigee [closest orbit to Earth] appears larger, brighter. At apogee, it's called a micromoon.

### BLOOD MOON
Refers to moon in total lunar eclipse, fully in Earth's shadow, holding reddish glow for up to two hours.

### BLUE MOON
The 13th full moon of the year rises every 30 months; if seeking bluish tint, find volcanic ash or forest fire.

### WET MOON
Crescent moon whose tips point upwards, also believed to be filled with rainwater in Hawaiian myth.

### HARVEST MOON
Full moon closest to autumnal equinox, historically vital to farmers harvesting crops late into the night.

### LONG NIGHT'S MOON
Final full moon of the year, also known as Cold Moon, occurs near winter solstice, hence "long night."

# BOOKS

**FROM THE EARTH TO THE MOON** by Jules Verne: A novel of post-Civil War America, in which the Baltimore Gun Club shoots three astronauts out of a cannon to the moon to settle a bet. The science was cutting edge for the time, except for the sticking point that the acceleration would have turned anyone actually riding the device into Jell-O. The novel inspired an opera, perhaps the first science fiction film [1902's *A Trip to the Moon*], and a ride at Disneyland Paris.

**OF A FIRE ON THE MOON** by Norman Mailer: A compendium of three long articles Mailer wrote for *Life* magazine in the immediate aftermath of the Apollo 11 mission. Much of it is written in an indulgent second-person style and shot through with references to astrology [Mailer refers to himself throughout as Aquarius]. It's heavy on philosophizing that 50 years later feels more than a bit dorm-room, an entertaining reminder that the moon landing happened at the very height of the Sixties counterculture.

**THE RIGHT STUFF** by Tom Wolfe: After being assigned to cover the last Apollo mission by Rolling Stone, the ever-competitive Wolfe was compelled to write an answer to Mailer's book on the space program. Rather than provide a historical account of the events of the space race, his curiosity was psychological: what sort of person agrees to be strapped to a rocket and blasted into space? The answer: stoic men with "the right stuff," a personality best exemplified by test pilot Chuck Yeager, the traces of which persist today in the laconic drawl of airline pilots all over the world.

**MAGNIFICENT DESOLATION** by Buzz Aldrin: The title refers to the lunar landscape, of course, but the primary subject is really the desolation of Aldrin's life after he became the second man to walk on the moon. After more than 40 years of single-minded striving, Aldrin was struck low by a profound emptiness upon returning to Earth. This memoir—his second—is an unflinching account of depression, adultery, alcoholism, and the slow path to rebuilding his life after he got sober in 1978.

# FILMS

**WOMAN IN THE MOON** [1929] A melodrama from science-fiction pioneer Fritz Lang, it was the first time many concepts in rocketry were depicted to a wide audience, such as multi-stage rockets and astronauts reclined to cope with the G-forces of launch. [*German rocket scientist, Hermann Oberth, considered one of the four most influential figures in early rocketry, served as a consultant.*]

.................................................................................

**2001: A SPACE ODYSSEY** [1968] Stanley Kubrick's outer-space masterpiece is two hours and forty-four minutes of more or less uninterrupted iconic cinema: the waltz of space stations set to Johann Strauss's "The Blue Danube," the murderous artificial intelligence, the glowing orbital baby— and, of course, the black monolith buried in the moon's Tycho crater.

.................................................................................

**FOR ALL MANKIND** [1989] In the late '70s, while conducting interviews with retired Apollo astronauts, director Al Reinert realized NASA was sitting on troves of unseen documentary footage. Reinert then spent a decade piecing it all together; all the Apollo missions as one single epic lasting 80 minutes, featuring a far-out soundtrack from Brian Eno, that feels at once both dated and wildly futuristic.

.................................................................................

**HIDDEN FIGURES** [2016] Before "computer" meant an IBM mainframe or iMac, it referred to a person who professionally performed complex math. NASA employed scores of computers, including black women like Katherine Johnson [played by Taraji P. Henson], who calculated the precise track of John Glenn's first orbit of the Earth and, later, the Apollo and Shuttle missions. The film is a fictionalized, but essentially accurate, account of the contributions of these women and the racism they faced.

.................................................................................

**APOLLO 11** [2019] In the spirit of Reinert and *For All Mankind*, Todd Douglas Miller takes archival footage from the Apollo 11 mission and crafts it into a verité, edge-of-your-seat, gasp-inducing narrative-free journey to the moon. It also features never before seen 70 mm film of the crowds watching the launch, as well as the armies of highly competent, mostly very young men in Houston's Mission Control who pulled off one of the more epic feats of humankind.

# DARK SKY SANCTUARIES

*The most remote [and often darkest] places in the world,
as determined by the International Dark Sky Association.*

**!AE!HAI KALAHARI HERITAGE PARK**
*Western South Africa*

**MASSACRE RIM WILDERNESS**
*Washoe County, Nevada, USA*

**AOTEA / GREAT BARRIER ISLAND**
*North Island, New Zealand*

**PITCAIRN ISLANDS**
*British Territory, South Pacific*

**GILA NATIONAL FOREST**
*Caltron County, New Mexico, USA*

**RAINBOW BRIDGE NATL. MONUMENT**
*Page, Arizona, USA*

**DEVILS RIVER / DEL NORTE**
*Del Rio, Texas, USA*

**STEWART ISLAND / RAKIURA**
*New Zealand*

**GABRIELA MISTRAL**
*Elqui Valley, Chile*

**VINDEX RANGE / THE JUMP-UP**
*Winton District, Australia*

---

# SPACE INSTITUTIONS

**ADLER PLANETARIUM**
*Chicago, IL*
America's first planetarium [1930] has three theaters and 8,000+ astro-artifact trove.

**GORNERGRAT PLANETARIUM**
*Zermatt, Switzerland*
Part semi-luxe hotel, part star-gazing trek, perched high in the snowy, moonlit Alps.

**NATIONAL AIR AND SPACE MUSEUM**
*Washington, D.C.*
The apex of Space Race objects, large and small, including Tranquility Base artifacts.

**KITT PEAK NATIONAL OBSERVATORY**
*Tucson, AZ*
Twenty-two telescopes in Sonoran Desert, 260 clear nights a year, open to public.

**SPACE CENTER HOUSTON**
*Houston, TX*
Starships from Mercury, Gemini and Apollo, moon rocks, full-sized Saturn V rocket.

**MEMORIAL MUSEUM OF COSMONAUTICS**
*Moscow, Russia*
Soviet space achievements are often overlooked; go here to see Gagarin's capsule.

# OBSERVATION

### PHASE SELECTION

Full moons are spectacular, but the best moon for viewing is a few days after first quarter. Follow the terminator line, the shadow edge, where lunar surface details are at highest contrast.

.............................................................................................

### PHOTOGRAPHY

To capture the moon, set camera's white balance for daylight, faster shutter speed and smaller aperture. The brightness is reflected sunlight after all. Place moon in context [structures, trees, even people] for added dimension.

.............................................................................................

### VIRTUAL MOON

Google Moon weaves together '60s-era moon maps and charts, created by the U.S. Air Force and Geological Survey, with Clementine multi-spectral lunar images, all to dramatically transportive effects.

---

# TOTAL ECLIPSES

Across human history and all geography, peoples have stood in fearful awe when the moon slides over the sun. For a few minutes, day becomes night. Nature reverses. Wars stop. The gods speak. In truth, this total alignment is a celestial miracle, the three bodies being precisely distant [the moon is 400x closer to Earth] and comparably sized [the sun is 400x larger than the moon]. Astronomers will tell you that, in the vastness of our Milky Way, this heavenly splendor of mathematical perfection is beyond distinctive, as in once-in-the-galaxy rare.

**DEC 14, 2020**
Argentina, Chile, Polynesia

.......................................................

**DEC 4, 2021**
Antarctica

.......................................................

**APR 8, 2024**
Mexico, United States, Canada

.......................................................

**AUG 12, 2026**
Arctic, Greenland, Iceland, Spain, Portugal

.......................................................

**AUG 2, 2027**
Morocco, Spain, Algeria, Libya, Egypt, Saudi Arabia, Somalia

.......................................................

**JULY 22, 2028**
Australia, New Zealand

.......................................................

**NOV 25, 2030**
Botswana, South Africa, Australia

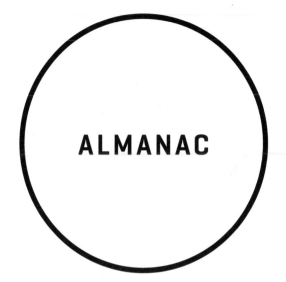

# ALMANAC

---

*A deep dive into the moon in human history through news clippings, timelines, artwork, literature, political speeches and other hearsay*

# SHAKESPEARE

*The moon: one of the Bard's most recurring motifs.*
*A brief sampling of its numerous appearances.*

It is the very error of the moon.
She comes more near the earth
than she was wont. And makes
men mad.
—*Othello*

How sweet the moonlight sleeps upon this bank!
Here will we sit and let the sounds of music
Creep in our ears: soft stillness and the night
Become the touches of sweet harmony.
—*The Merchant of Venice*

As the moon does, by wanting light to give:
But then renew I could not, like the moon;
There were no suns to borrow of.
—*Timon of Athens*

The pale-faced moon looks bloody on the earth
And lean-look'd prophets whisper fearful change;
—*Richard II*

O, swear not by the moon, the inconstant moon,
That monthly changes in her circled orb,
Lest that they love prove likewise variable.
—*Romeo and Juliet*

Four days will quickly steep themselves in night;
Four nights will quickly dream away the time;
And then the moon, like to a silver bow
New-bent in heaven, shall behold the night
Of our solemnities.
—*A Midsummer Night's Dream*

# LUNAR DEITIES OF NOTE

| | |
|---|---|
| Áine [*Celtic*] | Caretaker of agricultural pursuits, playtime in Lough Gur during a full moon |
| Auchimalgen [*Chilean*] | A protectress against evil spirits, aglow in crimson when death is imminent |
| Bahloo [*Aboriginal*] | Frequently found traipsing the Earth, oft-associated with serpents |
| Chandra [*Hinduism*] | Nightly sky streaker in a chariot pulled by 10 white horses/antelope |
| Chang'e [*Chinese*] | Goddess banished to the moon because she stole the pill of immortality from her husband |
| Coyolxauhqui [*Aztec*] | Decapitated by her god-of-war brother, head tossed to the heavens to anchor the night sky |
| Dewi Ratih [*Balinese*] | Chased—and sometimes consumed—by floating head of demon Kala Rau, an explainer for an eclipse |
| Igaluk [*Inuit*] | Incestuous, bloody cat-and-mouse game with sister led to forever placement in skies overhead |
| Ix Chel [*Mayan*] | "Lady Rainbow," a triple goddess with stewardship of the moon, water, weaving and fertility |
| Khonsu [*Egyptian*] | A "traveler," much depicted with crescent symbolism, possible representation in the Marvel Universe |
| Kuu [*Finnish*] | Daughter of air goddess, fashioned from the whites of duck eggs |
| Luna [*Roman*] | Shares lunar goddess duties with Diana and Juno, "two-horned queen of the stars" per Horace |
| Máni [*Norse*] | Along with sister Sól, ceaselessly chased across the heavens by a pair of wolves |
| Mawu [*Fon*] | Western Africa dweller, eclipses explained as lovemaking sessions with twin [sun deity] Lisa |
| Mayari [*Tagalog*] | One-eyed charmer, daughter of Bathala, totem of war, revolution, beauty and strength |

# CRUCIFIXION DARKNESS

*It was now about noon, and darkness came over the whole land until three in the afternoon, while the sun's light failed; and the curtain of the temple was torn in two. Then Jesus, crying with a loud voice, said, "Father, into your hands I commend my spirit."* — Luke 23: 44-46

According to the Gospels of Mark, Matthew, and Luke, the crucifixion of Jesus, on the eve of Passover, coincided with a curious darkening of the sun. Theologians have long debated the possible cause of such an event—perhaps a miracle or scribal error. Others, however, believe it may have been a total solar eclipse. Indeed, early manuscripts from the Gospel state directly that "the sun was in eclipse." But scientists, modeling the moon's history, have noted that, in this era, the only total eclipse visible in Jerusalem would have occurred in the year A.D. 29 on November 24 at 11:05 a.m. That eclipse also lasted less than two minutes—not, as noted in scripture, three hours.

## LUNACY

Since ancient times, full moons have been associated with odd or insane behavior, including sleepwalking, suicide, illegal activity, fits of violence and, of course, transforming into werewolves. Indeed, the words "lunacy" and "lunatic" come from the Roman goddess of the moon, Luna, who was said to ride her silver chariot across the dark sky each night. For thousands of years, doctors and mental health professionals believed in a strong connection between mania and the moon. Hippocrates, considered the father of modern medicine, wrote in the fifth century B.C. that "one who is seized with terror, fright and madness during the night is being visited by the goddess of the moon." In 18th-century England, people on trial for murder could campaign for a lighter sentence on grounds of lunacy if the crime occurred under a full moon; meanwhile, psychiatric patients at London's Bethlehem Hospital were shackled and flogged as a preventive measure during certain lunar phases. Even today, despite studies discrediting the hypothesis, some people think full moons make everyone a little loony.

Moon worship petroglyph at Honaki ruins in Sedona, Arizona.
Photo credit: Danita Delimont.

# WEREWOLVES

*Even a man who is pure in heart | and says his prayers by night | may become a wolf when the wolfbane blooms | and the autumn moon is bright.*
—poem recited numerous times in *The Wolf Man* [1941]

When Universal Pictures sought to add a werewolf to its stable of highly successful, horror film franchises of the 1930s—see *Dracula, Frankenstein, The Mummy,* et al.—studio heads turned to screenwriter Curt Siodmak to craft the character from some 2400 years' worth of lycanthropy myth. Smart choice. Siodmak's protagonist, an American who turns werewolf while visiting his ancestral home in Wales, was a new kind of movie monster: a brutal killer by night, while by day he suffered terrible guilt, fear he'd be killed himself, and a growing dread of his moonlit transformations. He also happened to be a whole new kind of werewolf. The tie to the lunar cycle, along with the prescriptive silver bullet—both staples of modern werewolf lore—were products not of Siodmak's extensive research but of his imagination. He simply invented them. "That four-liner has been attributed to 'Gypsy folklore'," Siodmak wrote in his autobiography, *Wolf Man's Maker.* "I made it up. That's how folk history is made."

---

**WAYS A FULL MOON CAUSES BAD LUCK:**

..............................................

When it falls on a Sunday

When you point at it over your shoulder

When you see the full moon in earthshine during a crescent— if you are a sailor, storm's a brewin'

**WAYS A FULL MOON CAUSES GOOD LUCK:**

..............................................

When you hold a moonstone in your mouth during it

On a Monday [Moon day]

When you see the first sliver of it clear from brush, over an open horizon

*From The Old Farmer's Almanac*

# THE FIRST MEN IN THE MOON

*By H. G. Wells, 1901*

As we saw it first it was the wildest and most desolate of scenes. We were in an enormous amphitheatre, a vast circular plain, the floor of the giant crater. Its cliff-like walls closed us in on every side. From the westward the light of the unseen sun fell upon them, reaching to the very foot of the cliff, and showed a disordered escarpment of drab and grayish rock, lined here and there with banks and crevices of snow. This was perhaps a dozen miles away, but at first no intervening atmosphere diminished in the slightest the minutely detailed brilliancy with which these things glared at us. They stood out clear and dazzling against a background of starry blackness that seemed to our earthly eyes rather a gloriously spangled velvet curtain than the spaciousness of the sky.

The eastward cliff was at first merely a starless selvedge to the starry dome. No rosy flush, no creeping pallor, announced the commencing day. Only the Corona, the Zodiacal light, a huge cone-shaped, luminous haze, pointing up towards the splendour of the morning star, warned us of the imminent nearness of the sun.

Whatever light was about us was reflected by the westward cliffs. It showed a huge undulating plain, cold and gray, a gray that deepened eastward into the absolute raven darkness of the cliff shadow. Innumerable rounded gray summits, ghostly hummocks, billows of snowy substance, stretching crest beyond crest into the remote obscurity, gave us our first inkling of the distance of the crater wall. These hummocks looked like snow. At the time I thought they were snow. But they were not—they were mounds and masses of frozen air.

So it was at first; and then, sudden, swift, and amazing, came the lunar day.

# THE HISTORY OF SCIENCE FICTION AND THE MOON

*A brief timeline of the genre's imaginings
of the moon—epic take-offs, colony collapse, and cheese*

1516......Ludovico Ariosto's Italian epic poem Orlando Furioso
becomes one of the first texts to imagine a flight to the moon.

1592 .....Japanese folktale The Tale of the Bamboo Cutter recounts a
Moon Princess's return home after years on Earth.

1638 .....After a deadly duel, the winner flees for the moon on the
back of a swan in Francis Godwin's *The Man in the Moone.*

1835......New York newspaper The Sun announces the discovery of man-
bats on the moon in what was dubbed the "Great Moon Hoax."

1901 .....H. G. Wells imagined moon jungles and snow [not to
mention life] in his novel *The First Men in the Moon.*

1928.....Writer Thea von Harbou's *Die Frau im Mond* inspired
rocket science and her filmmaker husband, Fritz Lang.

1950 .....Academy Award-winning *Destination Moon* deals with the
dangers of lunar landing and heralds the dawn of the Space Age.

1961 .....A tourist craft encounters a moonquake and liquefied moon
matter in Arthur C. Clarke's novel *A Fall of Moondust.*

1967.....Doctor Who visits the moon to use a weather controlling
Gravitron in the four-part TV series, *The Moonbase.*

1972 .....Sci-fi grandmaster Isaac Asimov envisages Earth-on-the-brink
and its lunar satellite in the novel *The Gods Themselves.*

1991 .....Aliens crash land on the moon and wage war with Earth in the
video game *Dead Moon.*

2000 ....An old-timer rides a Russian satellite to the moon in the Clint
Eastwood film *Space Cowboys.*

2005.....Laurie Anderson creates *The End of the Moon,* a ninety-
minute monologue based on her years as NASA's artist-in-
residence.

2011 .....Sci-fi horror film Apollo 18 uses "lost footage" to recount a
bloodstained final voyage to the moon.

2017.....Andy Weir's novel *Artemis* recounts an indolent moon-bound
criminal with too many smarts.

Chagall, Marc. *To the Moon.* 1953, Municipal Collection, Bielefeld.

# SONGS OF NOTE

**ALABAMA SONG** Brecht & Weill [1927]
*Oh moon of Alabama | We now must say goodbye | We've lost our good old mama | And must have whiskey...you know why*

**BLUE MOON** Rodgers & Hart [1934]
*Blue moon, you knew just what I was there for | You heard me saying a prayer for | Someone I really could care for*

**MOONLIGHT BECOMES YOU** Van Heusen & Burke [1942]
*If I say I love you | I want you to know | It's not just because there's moonlight | Although, moonlight becomes you so*

**BLUE MOON OF KENTUCKY** Bill Monroe [1946]
*Blue moon of Kentucky keep on shining | Shine on the one who's gone and left me bluer*

**OLD DEVIL MOON** Burton Lane [1947]
*I look at you and suddenly something in your eyes I see | Soon begins bewitching me | It's that old devil moon that you stole from the skies*

**I'M SO LONESOME I COULD CRY** Hank Williams [1949]
*The moon just went behind a cloud | To hide his face and cry*

**HAVANA MOON** Chuck Berry [1956]
*Havana moon, Havana moon | Me all alone with jug of rum | Me stand and wait for boat to come*

**MOON RIVER** Mercer & Mancini [1960]
*Moon river, wider than a mile | I'm crossing you in style some day*

**BLUE BAYOU** Melson & Orbison [1963]
*Oh, that girl of mine | by my side | The silver moon | and the evening tide | Oh, some sweet day | gonna take away | This hurtin' inside*

**I THINK IT'S GOING TO RAIN TODAY** Randy Newman [1964]
*Broken windows and empty hallways | A pale dead moon in the sky streaked with gray*

**NONE BUT THE RAIN** Townes Van Zandt [1969]
*None but the rain should cling to my bosom | None but the moon should hear my lonesome sigh*

**BOTH SIDES NOW** Joni Mitchell [1969]
*Moons and Junes and Ferris wheels | The dizzy dancing way you feel |*
*As every fairy tale comes real | I've looked at love that way*

**THE MOONBEAM SONG** Harry Nilsson [1971]
*Have you ever watched a moonbeam | As it slid across your windowpane |*
*Or struggled with a bit of rain | Or danced around a weathervane*

**PINK MOON** Nick Drake [1971]
*Saw it written and I saw it say | Pink moon is on its way | And none of*
*you stand so tall | Pink moon gonna get ye all*

**CAJUN MOON** J.J. Cale [1974]
*You're just like rain | To a love in bloom so | Shine on me oh, cajun moon*

**DRUNK ON THE MOON** Tom Waits [1974]
*And the moon's a silver slipper | It's pouring champagne stars*

**THE MOON IS A HARSH MISTRESS** Jimmy Webb [1974]
*The moon is a harsh mistress | She's so hard to call your own*

**SONG ABOUT THE MOON** Paul Simon [1981]
*So if you want to write a song about the heart | And its ever-longing for a*
*counterpart | Na na na na na na | Yeah yeah yeah | Write a song about the moon*

**BLUE MOON WITH HEARTACHE** Rosanne Cash [1981]
*Blue moon out my window | Guess this means goodnight*

**KIKO AND THE LAVENDER MOON** Hidalgo & Perez [1982]
*Kiko and the lavender moon | Out dreaming 'bout green shoes |*
*Haircuts and cake*

**LET'S DANCE** David Bowie [1982]
*Let's dance | Put on your red shoes and dance the blues | Let's sway |*
*Under the moonlight, this serious moonlight*

**CLOSING TIME** Leonard Cohen [1992]
*And the moon is swimming naked | And the summer night is fragrant |*
*With a mighty expectation of relief*

**RESPIRATION** Mos Def and Talib Kweli [1998]
*The new moon rode high in the crown of the metropolis |*
*Shining, like who on top of this?*

## THE MOON IN FILM

Georges Méliès's 1902 classic, *Le Voyage dans la Lune,* was one of the most influential and earliest examples of science fiction film. The plot follows a band of astronomers who travel to the Moon in a canon-fired capsule [which famously lands in the eye of the Moon's eye] and find a band of lunar inhabitants called Selenites, after the Greek goddess of the Moon. A silent picture running around 16 minutes, the film was internationally popular, especially in the United States where it toured through vaudeville theaters and appeared at fairs. Decades after the film's release, film pioneer and projection developer Jean Acme LeRoy sent a series of written questions to Méliès. The questions were lost, but Méliès's responses remain. A Frenchman, he nevertheless responded to the "questionary" in English, with idiosyncratic spelling and grammar.

*The idea of "Trip to the Moon" came to me from the book of Jules Verne, entitled: "From the earth to the moon and round the moon." In this work the human people could not attain the moon, turn round it, and came back to earth, having, in fact, missed their trip. I then imagined, in using the process of Jules Verne, [gun and shell] to attain the moon, in order to be able to compose a number of original and amusing fairy pictures outside and inside the moon, and to show some monsters, inhabitants of the moon in adding one or two artistical effects [women represent-ing stars, comets, etc.] [snow effect, bottom of the sea, etc]... —Georges Méliès, 1930*

# MOON MYTHS OF NOTE

### CHINA

The child of Yin and Yang, P'an Ku, was born at the very beginning of everything and spent 18,000 years creating the stars, the Earth, the Moon, the sky, the whole universe. Only, he forgot something. The Sun and Moon weren't in the sky, because P'an Ku left them in the Han Sea, and the world remained in darkness until the Buddah intervened and had him make things right by writing the name of the Sun and Moon on his hands, going into the sea and performing a ritual seven times until they rose up and took their place in the sky and divided the seven days a week into day and night.

### GREECE

The Moon goddess Selene fell in love with Endymion, a shepherd on Mount Latmus, while he was dozing in a cave. Selene asked Zeus to put Endymion into an endless sleep, so she could be with him forever.

### MAORI

There is a lake in the land above the heavens, a place known as the land of the water life and the gods. Here, in this lake called Ka-ne, in the water life, the living return from the dead. And so the Moon goes into the water each day, to be returned to her path in the sky at night.

### AKAN (WEST AFRICA)

Anasi, the spider god, saw a bright light in the forest, and went to get it to reward his six handsome and loyal sons. While carrying the light back to his village, he wondered how he would divide it between them—there was but one light, and six sons to share it. The wise owl swooped down and said he would take this light and fly to a place in the sky and put the light there for them all to enjoy equally, and there it remains today, the Moon.

### CADDO

For thousands of years throughout Native North America, Caddo priests called tsah neeshi [Mr. Moon] were empowered by a'a caddiayo [Father Above Chief] and their hero neesh [Moon]. Post-harvest celebrations followed the September New Moon, and over the centuries, "moon" has woven in language to name ringed ceremony centers and crescent-shaped peyote altars.

# POETRY

### ABOVE THE DOCK
T.E. Hulme

Above the quiet dock in midnight,
Tangled in the tall mast's corded height,
Hangs the moon. What seemed so far away
Is but a child's balloon, forgotten after play.

...............................................................................................

### FULL MOON AND LITTLE FRIEDA
Ted Hughes

A cool small evening shrunk to a dog bark and the clank of a bucket -
And you listening.
A spider's web, tense for the dew's touch.
A pail lifted, still and brimming - mirror
To tempt a first star to a tremor.

Cows are going home in the lane there, looping the hedges with their warm
wreaths of breath -
A dark river of blood, many boulders,
Balancing unspilled milk.
'Moon!' you cry suddenly, 'Moon! Moon!'

The moon has stepped back like an artist gazing amazed at a work
That points at him amazed.

...............................................................................................

### THE MOON
Sappho

The stars about the lovely moon
Fade back and vanish very soon,
When, round and full, her silver face
Swims into sight, and lights all space

## THE MOON WAS BUT A CHIN OF GOLD
Emily Dickinson

The moon was but a chin of gold
 A night or two ago,
And now she turns her perfect face
 Upon the world below.

Her forehead is of amplest blond;
 Her cheek like beryl stone;
Her eye unto the summer dew
 The likest I have known.

Her lips of amber never part;
 But what must be the smile
Upon her friend she could bestow
 Were such her silver will!

And what a privilege to be
 But the remotest star!
For certainly her way might pass
 Beside your twinkling door.

Her bonnet is the firmament,
 The universe her shoe,
The stars the trinkets at her belt,
 Her dimities of blue.

# SPIRITUALITY AND THE MOON

### RALPH WALDO EMERSON

"The man, who has seen the rising moon break out of the clouds at midnight, has been present like an archangel at the creation of light and of the world." *The Essays of Ralph Waldo Emerson*

### SIMONE WEIL

"The full moon is something perfect which the very next day will no longer be visible. The moon is an object which can be contemplated face to face, unlike the sun. The moon is the last thing to be beheld by the man who has emerged from the Cave described in Plato, immediately before being rendered capable of casting a look—necessarily a fugitive one—at the sun. In other words, according to the Symposium, it represents the beauty of God." *The Notebooks of Simone Weil*

### THOMAS MERTON

"What can we gain by sailing to the moon if we are not able to cross the abyss that separates us from ourselves? This is the most important of all voyages of discovery, and without it all the rest are not only useless but disastrous." *The Wisdom of the Desert: Sayings from the Desert Fathers of the Fourth Century*

### MARILYNNE ROBINSON

"The moon looks wonderful in this warm evening light, just as a candle flame looks beautiful in the light of morning. Light within light. It seems like a metaphor for something. So much does. Ralph Waldo Emerson is excellent on this point. It seems to me to be a metaphor for the human soul, the singular light within the great general light of existence." *Gilead*

### JORGE LUIS BORGES

"There is so much loneliness in that gold. The moon of every night is not the moon that the first Adam saw. The centuries of human wakefulness have left it brimming with ancient tears. Look at her. She is your mirror." *The Moon*

### MAHATMA GANDHI

"When I admire the wonders of a sunset or the beauty of the moon, my soul expands in the worship of the creator."

Moon model prepared by Johann Friedrich Julius Schmidt in 1898.
Made of 116 sections of plaster on wood and metal.

# KENNEDY

*When John F. Kennedy became president in 1961, the world believed the U.S. was losing the Space Race. The Soviet Union had already put a satellite, a dog and a man in space. Kennedy, convinced that Americans needed a win, proposed a manned mission to the moon, and he used a sunny day at Rice University, before a crowd of 40,000, to convince the country that such a feat was possible—and necessary.*

HOUSTON, TEXAS
September 12, 1962

"Those who came before us made certain that this country rode the first waves of the industrial revolutions, the first waves of modern invention, and the first wave of nuclear power, and this generation does not intend to founder in the backwash of the coming age of space. We mean to be a part of it; we mean to lead it. For the eyes of the world now look into space, to the moon and to the planets beyond, and we have vowed that we shall not see it governed by a hostile flag of conquest, but by a banner of freedom and peace. We have vowed that we shall not see space filled with weapons of mass destruction, but with instruments of knowledge and understanding.

Yet the vows of this nation can only be fulfilled if we in this nation are first, and, therefore, we intend to be first. In short, our leadership in science and in industry, our hopes for peace and security, our obligations to ourselves as well as others, all require us to make this effort, to solve these mysteries, to solve them for the good of all men, and to become the world's leading space-faring nation.

We set sail on this new sea because there is new knowledge to be gained, and new rights to be won, and they must be won and used for the progress of all people. For space science, like nuclear science and all technology, has no conscience of its own. Whether it will become a force for good or ill depends on man, and only if the United States occupies a position of pre-eminence can we help decide whether this new ocean will be a sea of peace or a new terrifying theater of war. I do not say that we should or will go unprotected against the hostile misuse of space any more than we go unprotected against the hostile use of land or sea, but I do say that space can be explored and mastered without feeding the fires of war, without repeating the mistakes that man has made in extending his writ around this globe of ours.

There is no strife, no prejudice, no national conflict in outer space as yet. Its hazards are hostile to us all. Its conquest deserves the best of all mankind, and its opportunity for peaceful cooperation may never come again. But why, some say, the moon? Why choose this as our goal? And they may well ask why climb the highest mountain? Why, 35 years ago, fly the Atlantic? Why does Rice play Texas?

We choose to go to the moon. We choose to go to the moon in this decade and do the other things, not because they are easy, but because they are hard, because that goal will serve to organize and measure the best of our energies and skills, because that challenge is one that we are willing to accept, one we are unwilling to postpone, and one which we intend to win, and the others, too."

---

## MERCURY 7

**SCOTT CARPENTER** *Second to orbit Earth, his splashdown 250 miles off course called "joyriding" by* [*some*] *NASA colleagues*

**GORDON COOPER** *Speedboat racer and UFO believer, Cooper was first to sleep in space and last to do solo orbital mission*

**JOHN GLENN** *Famously circled Earth three times, later served Ohio as Senator for 25 years and went back to space at age 77*

**GUS GRISSOM** *Second to fly in space, escaping near-tragedy when capsule sunk in Atlantic; died in Apollo 1 pre-launch fire*

**WALLY SCHIRRA** *Flew in Mercury, Gemini and Apollo missions to space, co-anchored moon landings on CBS with Walter Cronkite*

**ALAN SHEPARD** *First American in space and only Mercury 7 member to walk on the moon; hit two golf balls from the surface*

**DEKE SLAYTON** *Grounded due to heart issues, directed Flight Crew Operations for all Apollo missions, finally flew in 1973*

# APOLLO PATCHES

**APOLLO 1**
JANUARY 27, 1967

**APOLLO 7**
OCTOBER 11, 1968

**APOLLO 8**
DECEMBER 21, 1968

**APOLLO 9**
MARCH 3 1969

**APOLLO 10**
MAY 18, 1969

**APOLLO 11**
JULY 16, 1969

**APOLLO 12**
NOVEMBER 14, 1969

**APOLLO 13**
APRIL 11, 1970

**APOLLO 14**
JANURARY 31, 1971

**APOLLO 15**
JULY 26, 1971

**APOLLO 16**
APRIL 16, 1972

**APOLLO 17**
DECEMBER 7, 1972

## APOLLO CONTRIBUTORS OF NOTE

### WERNHER VON BRAUN
Former Nazi designer recruited by US government after WWII, led team that built Saturn V, the powerful rocket that carried Apollo astronauts to the moon.

### ELEANOR FORAKER
A senior seamstress at the International Latex Corporation, a company better known for Playtex underwear, she helped lead the team hand-sewing the Apollo spacesuits.

### ROBERT GILRUTH
The boss of many better-known Apollo figures, he established and ran the Manned Spacecraft Center ["Houston"].

### MARGARET HAMILTON
The MIT scientist who led the team that developed the Apollo spacecraft's in-flight software.

### DEKE SLAYTON
The man who decided Neil Armstrong would be the first to walk on the moon. After he was pulled from the Mercury program because of an irregular heartbeat, became Director of Flight Crew operations, NASA's astronaut boss. [Finally traveled to space in 1975 at age 51.]

### KATHERINE JOHNSON
A master of the complex math behind orbits and trajectories, she did the calculations for the rendezvous of the lunar lander with the command module as they orbited the moon.

### YURI KONDRATYUK
The Ukranian engineer who, as a purely theoretical exercise, first proposed the lunar orbit rendezvous plan eventually used by the Apollo program. Killed during WWII while fighting for the USSR.

### GENE KRANZ
The Chief Flight Director for Mercury and Apollo, is now best remembered for directing the operation to recover Apollo 13.

### HOMER NEWELL
As the director of all of the science that the Apollo missions carried out, he chose the final design of the scientific instruments left behind by Apollo scientists.

### JAMES WEBB
Administrator of NASA during almost all of the Space Race. By his own admission not knowledgeable about rocketry or space, he was a bureaucratic master who kept programs funded and on time.

*Mercury-Redstone 3* ...........................................*May 5*, 1961/*May 5*,1961
The first human-crewed U.S. spaceflight, piloted by Alan Shepard. It
remained in sub-orbital flight for 15 minutes before reentry.

*Mercury-Atlas 6* ...............................*Feb.* 20, 1962/*Feb.* 20, 1962
John Glenn became the first American to orbit the Earth. After nearly five
hours and three full orbits, Glenn splashed down in the Atlantic Ocean.

*Apollo 11* ........................................ *Jul.* 16, 1969/*Jul.* 24, 1969
The first humans—Neil Armstrong and Buzz Aldrin—landed on the
Moon, ending the Space Race and fulfilling a worldwide goal.

*Apollo 13* ........................................*April 11*, 1970/*April 17*, 1970
Intended to be the third lunar landing, the mission was aborted after a
cascade of critical failures. On-the-fly fixes brought everyone home safe.

*Voyager 1* ...............................................................*Sep. 5*, 1977—
The probe flew by Jupiter and Saturn and captured a photo of the
entire solar system. In 2012, it crossed into interstellar space—a first.

*STS-51-L* ....................................*Jan.* 28, 1986—*Jan.* 28, 1986
The mission ended 73 seconds after liftoff when a failed rocket booster
destroyed the Challenger space shuttle, killing all seven crew members.

*Hubble* .......................................................................*April 24*, 1990—
One of the largest and most versatile space telescopes, it changed the
way humans understand the cosmos.

*International Space Station* ............................................ *Nov.* 20, 1998—
A joint project with Canada, Europe, Japan, and Russia, the low-orbit space
station has been habited since 2000 and is home to countless experiments.

*STS-107* .............................................................*Jan.* 16, 2003/*Feb.* 1, 2003
The Space Shuttle program's 113th flight, the mission lasted nearly 16 days
until its space shuttle, the Columbia, broke up during reentry, killing the crew.

*Kepler* ...........................................................................*Mar.* 7, 2009—
The space telescope—and its successor, K2—have monitored more
than 150,000 stars in search of other Earth-like exoplanets.

NASA Launch Control Center for Apollo 11.
Kennedy Space Center, Florida. Photo credit: NASA.

# TRANQUILITY BASE

*With the world watching, CBS anchorman Walter Cronkite and
astronaut Wally Schirra led coverage of the Apollo 11 moon landing.*

CAPCOM: Thirty seconds.

EAGLE: Contact light. OK, engine stopped, descent engine command override off.

SCHIRRA: We're home.

CRONKITE: Man on the moon!

CAPCOM: We copy you now, Eagle.

SCHIRRA: Oh, geez.

EAGLE: Houston, Tranquility Base here. The Eagle has landed.

CAPCOM: Roger, Tranquility. We copy you on the ground. You've got a bunch of guys about to turn blue. We're breathing again. Thanks a lot.

CRONKITE: [*Laughing*] Oh, boy!

TRANQUILITY: Thank you.

CAPCOM: You're looking good here.

CRONKITE: Whew! Boy!
[*Schirra wipes eye. Cronkite takes off glasses, smiles, rubs hands together.*]

TRANQUILITY: OK, we're going to be busy for a minute.

CRONKITE: Wally, say something. I'm speechless.

SCHIRRA: I'm just trying to hold onto my breath. That is really something. Kinda nice to be aboard on this one, isn't it?

CRONKITE: You know, we've been wondering what, what this guy Armstrong and Aldrin would say when they set foot on the moon, which comes a little bit later now. Just to hear them do it, we're left with absolutely dry mouths and speechless.

CAPCOM: Roger, Eagle. And you're stay for T-1. Over. You're stay for T-1...

TRANQUILITY: Roger. We're stay for T-1.

CAPCOM: Roger. And we see you getting the ox.

CRONKITE: That's a great simulation that we see here, if you can see it.

SCHIRRA: That little fly-speck is supposed to be the LM.

CRONKITE: They must be in perfect condition, upright, we've heard no complaints about their position.

SCHIRRA: Just a little dust, even that concern was erased.

CRONKITE: Wow! And there they sit on the moon. Just exactly nominal isn't it, exactly with the flight plan, all the way down. Man planning this thing on the surface on the surface of the Earth. My God, we've done it.

## PAGE ONE

*The New York Times*
New York City

MEN WALK ON MOON

Men have landed and walked on the moon. Two Americans, astronauts of Apollo 11, steered their fragile four-legged lunar module safely and smoothly to the historic landing yesterday at 4:17:40 P.M., Eastern daylight time.

*Il Messaggero*
Rome

LUNA PRIMA PASSO

They came down! Armstrong and Aldrin, with the help of Collins waiting for them in lunar orbit, made it. We do not believe we can overstate this greatest of achievements, the most fantastic in human history.

*Evening Standard*
London

MOONFALL!

After a fantastic voyage, a fitful sleep. Man conquered the moon today, watched on live television by a fifth of the world population—600 million people.

*Pravda*
Moscow

THE FIRST LUNAR EXPEDITION

We highly respect the American "Apollo 11" space flight, in which two people—Neil Armstrong and Edwin Aldrin—set foot on the lunar surface for the first time... It is impossible not to admire the courage and endurance of the cosmonauts who boldly met the unknown.

*Wapakoneta Daily News*
Wapakoneta, Ohio

NEIL STEPS ON THE MOON

The mother of Apollo 11 astronaut Neil Armstrong said today she was concerned her son would "sink in too deeply" when he set foot on the moon. "I was worried that the moon might be too soft and that he would sink in too deeply," Mrs. Viola Armstrong said. "But I'm so thankful they got there safely."

# "MOON PROBE LAUDABLE – BUT BLACKS NEED HELP"

*Jet Magazine*
By Simeon Booker

July 31, 1969

Landing an astronaut on the moon has more priority in America than putting a black man on his feet, in a job, or a poor family on a decent diet. This space accomplishment at a cost of billions of dollars will receive coast to coast acclaim and international attention.

But as a black Washington correspondent, I see this week as a crucial period in history. There will be headlines and hours of radio and television time on the day to day activity. President Nixon invited the president of his alma mater, Whittier College, to speak at the White House religious service on "the Meaning of The Man on The Moon." Meanwhile what of the man in the street—in poverty stricken Appalachia, Watts, and Harlem. He wished the astronauts well and marvels at their courage.

But he also wonders if the powers of science and technology will ever focus in such a fashion on his problems. Thanks to modern communications, even the simplest ghetto dweller knows that the American space program and its counterpart in the Soviet Union are almost as political in their motives as they are scientific.

And while the victims of poverty watch the space race with awe, we wonder how long it will be before the hypnosis of a moon flight wears off and the victims of poverty realize that they are still hungry. Perhaps the presence of the mule train of the Poor People's Campaign at Cape Kennedy will remind some people that their NASA tax dollars might best be spent in other ways.

Sometime, somehow, we Americans—and the Russians as well—must think about making the earth a better place to live. To escape to the moon is no answer for any of us—black, white, brown or yellow.

## ALTERNATIVE SPEECH

*Days before Apollo 11 was scheduled to touch down on the moon,
speechwriter William Safire drafted a memo to White House
Chief of Staff H.R. Haldeman, containing a statement to be read by
President Nixon if astronauts Neil Armstrong and Buzz Aldrin
became stranded on the lunar surface, never to return home. Safire
also noted that a clergyman should adopt the same procedure as a
burial at sea, commending their souls to "the deepest of the deep."*

### IN EVENT OF MOON DISASTER:

Fate has ordained that the men who went to the moon to explore in peace will stay on the moon to rest in peace.

These brave men, Neil Armstrong and Edwin Aldrin, know that there is no hope for their recovery. But they also know that there is hope for mankind in their sacrifice.

These two men are laying down their lives in mankind's most noble goal: the search for truth and understanding.

They will be mourned by their families and friends; they will be mourned by the nation; they will be mourned by the people of the world; they will be mourned by a Mother Earth that dared send two of her sons into the unknown.

In their exploration, they stirred the people of the world to feel as one; in their sacrifice, they bind more tightly the brotherhood of man.

In ancient days, men looked at the stars and saw their heroes in the constellations. In modern times, we do much the same, but our heroes are epic men of flesh and blood.

Others will follow, and surely find their way home. Man's search will not be denied. But these men were the first, and they will remain the foremost in our hearts.

For every human being who looks up at the moon in the nights to come will know that there is some corner of another world that is forever mankind.

Apollo 11 lunar module pilot Buzz Aldrin's bootprint at
Tranquility Base. Photo credit: Buzz Aldrin.

# THE SPACE RACE IN POP CULTURE

*Pop icons and masterworks from the height of the Space Race*

| DEBUT | WORK |
| --- | --- |

1949................. *Googie architecture*
John Lautner designed L.A. coffee shops, "Googies,"
with curvaceous look for the Space Age

1951................. *The Day the Earth Stood Still*
Klaatu and Gort come to Earth to deliver an
important message, with theremin accompaniment

1951................. *"Fly Me to the Moon"*
Frank Sinatra's 1964 version was played during
Apollo 10's moon orbit

1952................. *"Man Will Conquer Space Soon!"*
Chesley Bonestell, "father of modern space art,"
inspired the U.S. space program with iconic
illustrations

1954................. *Space age pop*
"Out there" listening with seductive moods and
techno-effects inspired by the stars

1955................. *"Man in Space"*
Walt Disney's "Science Feature from Tomorrowland"
may have prompted U.S. satellite program

1957................. *Tang*
NASA food favorite, the powdered drink was picked
for John Glenn's zero-G experiments

1958................. *"Sputnicks and Mutnicks"*
An ode to Russia's space pooches by Ray Anderson
and the Home Folks

1961................. *Theme Building*
Flying saucer building with four legs landed at LAX,
dedicated by Lyndon B. Johnson

1962 .................. *The Jetsons*
George Jetson and family flew through prime time with videophones and a robo-maid

1962 .................. *Barbarella*
Sexually-liberated, intergalactic Barbarella rocked the comic book world, then the big screen

1962 .................. *Space Needle*
Built for the Seattle World's Fair, it could have been a ballon but ended up a saucer

1964 .................. *Courrège's Spring collection*
Godfather of space age fashion André Courrèges celebrated PVC goggles, helmets, and flat boots

1965 .................. *Lost in Space*
TV's Robinson family drift through space with no end in sight

1966 .................. *Star Trek*
NBC introduced the world to the 2260s, Captain Kirk, and "Live long and prosper"

1967 .................. *The Amazing Urbmobile*
City planners fantasized about Space Age-influenced commuter travel that might have eliminated traffic

1968 .................. *2001: A Space Odyssey*
Stanley Kubrick won a space race with NASA to release his film before the moon landing

1968 .................. *Operating Manual For Spaceship Earth*
Buckminster Fuller declared in his architectural manifesto: "We are all astronauts"

1969 .................. *Stoned Moon series*
"Space artist" Robert Rauschenberg remembered Apollo 11 in lithographic series

1969 .................. *Space Oddity*
Ziggy jammed good with Weird and Gilly and the spiders from Mars

# WHITEY ON THE MOON

*by Gil Scott-Heron*

A rat done bit my sister Nell.
[with Whitey on the moon]
Her face and arms began to swell.
[and Whitey's on the moon]
I can't pay no doctor bill.
[but Whitey's on the moon]
Ten years from now I'll be payin' still.
[while Whitey's on the moon]
The man jus' upped my rent las' night.
['cause Whitey's on the moon]
No hot water, no toilets, no lights.
[but Whitey's on the moon]
I wonder why he's uppi' me?
['cause Whitey's on the moon?]
I was already payin' 'im fifty a week.
[with Whitey on the moon]
Taxes takin' my whole damn check,
Junkies makin' me a nervous wreck,
The price of food is goin' up,
An' as if all that shit wasn't enough:
A rat done bit my sister Nell.
[with Whitey on the moon]
Her face an' arm began to swell.
[but Whitey's on the moon]
Was all that money I made las' year
[for Whitey on the moon?]
How come there ain't no money here?
[Hm! Whitey's on the moon]
Y'know I jus' 'bout had my fill
[of Whitey on the moon]
I think I'll sen' these doctor bills,
Airmail special
[to Whitey on the moon]

## JOSEPH CAMPBELL

*The Moon Walk—the Outward Journey*

The only really adequate public comment on the occasion of the first moon walk that I have found reported in the world press was the exclamation of an Italian poet, Giuseppe Ungaretti, published in the picture magazine *Epoca*. In its vivid issue of July 27, 1969, we see a photo of this white-haired old gentleman pointing in rapture to his television screen, and in the caption beneath are his thrilling words: *Questaèuna notte diversa da ogni ultra notte del mondo*. For indeed that was "a different night from all other nights of the world!" Who will ever in his days forget the spell of the incredible hour, July 20, 1969, when our television sets brought directly into our living rooms the image of that strange craft up there and Neil Armstrong's booted foot coming down, feeling cautiously its way - to leave on the soil of that soaring satellite of earth the first impress ever of life? And then, as though immediately at home there, two astronauts in their space suits were to be seen moving about in a dream-landscape, performing their assigned tasks, setting up the American flag, assembling pieces of equipment, loping strangely but easily back and forth: their pictures brought to us, by the way, through two hundred and thirty-eight thousand miles of empty space by that other modern miracle [also now being taken for granted], the television set in our living room. "All humanity," Buckminster Fuller once said, in prophecy of these transforming forces working now upon our senses, "is about to be born in an entirely new relationship to the universe." To predict what the imagery of the poetry of man's future is to be, is today, of course, impossible. However, those same three astronauts, when coming down, gave voice to a couple of suggestions. Having soared beyond thought into boundless space, circled many times the arid moon, and begun their long return: how welcome a sight, they said, was the beauty of their goal, this planet Earth, "like an oasis in the desert of infinite space!" Now there is a telling image: this earth, the one oasis in all space, an extraordinary kind of sacred grove, as it were, set apart for the rituals of life; and not simply one part or section of this earth, but the entire globe now a sanctuary, a set-apart Blessed Place. Moreover, we have all now seen for ourselves how very small is our heaven-born earth, and how perilous our position on the surface of its whirling, luminously beautiful orb.

## SOVIET ATHLETES CHEER LANDING

*July 21, 1969*

ANAHEIM [AP]—Eighty members of the Russian track and field team cheered, slapped each other on the back and shook hands with those near them yesterday as they watched American astronauts land on the moon. The team, which competed in an international meet in Los Angeles Friday and Saturday, spent yesterday at Disneyland, where a giant screen in the Tomorrowland section showed televised coverage of the landing. "The Russians were just as much there as anybody else," a Disneyland official said. "There were tears and signs of sustained joy." Some of the team members took the Disneyland simulated trip to the moon, a nine-minute demonstration sponsored by the McDonnell Douglas Corp., an Apollo contractor. The landing wasn't broadcast in the U.S.S.R.

## NASA ART PROGRAM

Established in 1962, artists were given unprecedented access to NASA sites, launches and personnel, and allowed to create with little to no guidance. Since the launch, more than 3,000 works have been created to date by an illustrious and group, a few listed below:

| | |
|---|---|
| Norman Rockwell | Andy Warhol |
| Annie Leibovitz | Robert Rauschenberg |
| James Wythe | Ellen Levy |
| Sara Larkin | Henry Casselli |
| Robert A.M. Stephens | Peter Hurd |
| Robert McCall | Lamar Dodd |
| Mitchell Jamieson | Theodore Hancock |

*Artist Paul Calle not only sketched the Apollo 11 crew, he was later commissioned to create the commemorative 10-cent stamp by the USPS.*

Schultz, Charles. *Peanuts*. March 14, 1969. Reprinted with permission.

# BEYOND NASA

1957 ... The Soviet Union successfully launches Sputnik 1, the first artificial Earth satellite.

1957 ... Soviet dog Laika orbits the Earth, the first animal to do so, but dies within hours.

1959 ... Luna 2 of the U.S.S.R. becomes the first spacecraft to reach the lunar surface. Luna 3 captures images of the far side of the moon.

1961 ... Russian cosmonaut Yuri Gagarin makes 108-minute orbital flight aboard Vostok 1, becoming the first man in space.

1966 ... Luna 9 becomes first spacecraft to land successfully on the moon.

1969 ... While Apollo 11 astronauts finish their first moonwalk, the U.S.S.R.'s unmanned Luna 15 begins its descent to the lunar surface.

1976 ... The U.S.S.R. Luna program ends with Luna 24, which returned six ounces of lunar regolith.

1990 ... The first lunar probe since Luna 24, Japan's Hiten enters a three-year orbit around the moon.

1998 ... Japan aims Nozomi probe for Mars using lunar swingbys, but the mission is terminated after electrical failures.

2003 ... European Space Agency launches SMART-1 lunar satellite, the first European moon mission.

2007 .. Japan sends three more orbiters to the moon. China sends its first lunar spacecraft.

2008 .. India's Chandrayaan-1 lifts off, discovers water on the moon.

2013 ... Yutu, a lunar rover, successfully lands on the moon after being deployed by China's Chang'e 3.

2014 ... LuxSpace, a Luxembourg-based contractor, launches the Manfred Memorial Moon Mission.

2019 ... Space IL's Beresheet, the first Israeli and first privately funded moon lander mission, crashes into the lunar surface.

## APOLLO 15 SURVIVAL KIT

### CONTENTS

SURVIVAL RADIO

SURVIVAL LIGHT

ASSEMBLY

DESALTER KITS

MACHETE

SUNGLASSES

WATER CANISTERS

SUN LOTION

BLANKET

POCKET KNIFE

NETTING

FOAM PADS

*Manufacturer:*
B. Welson & Co.

*Origin:*
Hartford, Conn.

*Year:*
circa 1970

*Materials:*
Armalon, beta cloth,
brass, foam, glass, plastic,
steel, Velcro

*Dimensions:*
53.3 x 30.5 x 17.8 cm

## JOHN GLENN'S CAMERA

Most NASA equipment takes thousands, if not millions, of dollars and countless hours of testing. The first human-shot, color photos from space, however, were taken on a handheld camera that John Glenn found at a Cocoa Beach drugstore before his 1962 Friendship 7 flight.

*Make:*
Ansco Autoset

*Manufacturer:*
Minolta

*Origin:*
Japan, U.S.A.

*Year:*
1962

*Film format:*
35 mm

*Materials:*
Glass, metal, plastic,
quartz, Velcro

*Dimensions:*
13.5 × 7.5 × 24.5 cm

*List price:*
$52

# WALTER CRONKITE

July 24, 1969
*CBS Evening News*

Well, man's dream and a nation's pledge have now been fulfilled. The lunar age has begun. And with it, mankind's march outward into that endless sky from this small planet circling an insignificant star in a minor solar system on the fringe of a seemingly infinite universe. The path ahead will be long; it's going to be arduous; it's going to be pretty doggone costly. We may hope, but we should not believe, in the excitement of today, that the next trip or the ones to follow are going to be particularly easy. But we have begun with 'a small step for a man, a giant leap for mankind,' in Armstrong's unforgettable words. In these eight days of the Apollo 11 mission the world was witness to not only the triumph of technology, but to the strength of man's resolve and the persistence of his imagination. Through all times the moon has endured out there, pale and distant, determining the tides and tugging at the heart, a symbol, a beacon, a goal. Now man has prevailed. He's landed on the moon, he's stabbed into its crust; he's stolen some of its soil to bring back in a tiny treasure ship to perhaps unlock some of its secrets. The date's now indelible. It's going to be remembered as long as man survives—July 20, 1969—the day a man reached and walked on the moon. The least of us is improved by the things done by the best of us. Armstrong, Aldrin and Collins are the best of us, and they've led us further and higher than we ever imagined we were likely to go.

Photo known as *Earthrise*, taken by Apollo 8 astronaut William Anders, on December 24, 1968. NASA.

# INTERVIEWS

Twelve conversations with men and women about moon walking, planetary geology, space travel, Mission Control protocol, presidential politics, the early days of NASA, eclipse chasing and more

# ANDREA JONES

*PLANETARY GEOLOGIST*

**HALF OF MY** job at NASA is moon related, particularly given International Observe the Moon Night.

**IT STARTED 10** years ago, celebrating the arrival of the Lunar Reconnaissance Orbiter and LCROSS missions. We went to the moon with these missions and decided to have a big party to celebrate.

**THE NIGHT IS** a celebration of celestial observing and personal and cultural connections with the moon. Everybody everywhere can see the moon at some point. It's universal.

**I USED TO** teach teachers about the moon, and I would open up all the workshops with a question: what is your favorite memory with the moon in it?

**"I REMEMBER MY** child pointing to the moon," one would say. Another got married on July 20 so they would remember their anniversary because that's the moon landing. Another described a kayak trip beneath the moon.

**I HEAR A** lot from military families, a lot of personnel, from Afghanistan, the Middle East, East Asia, how they're on a call and say, look up at the moon, because at both ends of the call they can all see the same moon.

**IT'S SO POWERFUL,** such a unifying thing we all have.

.......................................

**OUR CULTURES HAVE** all been influenced by the moon. Look at our language: "Monday!" "Lunatic." It all ties back.

.......................................

**WE HAD ONE** host who wrote us about setting up in a parking lot on a roadside in Vermont and this motorcycle group came by and looked through telescopes for the first time ever. They were just blown away, all these tattooed guys.

.......................................

**EVEN IN THE** most light polluted skies, there's still stuff to see with the naked eye, like the Tycho crater. I think that the most obvious thing to me when I look up at the moon are the bright areas and dark areas.

.......................................

**THE LIGHTER AREAS** are the highlands, the anorthosite, and the darker areas are basalt, and those are mari basalt, the kind of rock you see in Hawaii or Craters of the Moon in Idaho.

.......................................

**LOOKING ALONG THE** terminator—my favorite planetary term, it's the line between day and night—that's where the best observing is. A group of lunar observers get excited about the moon not when it's a full moon, but when the

shadows are the best, a quarter moon.

.......................................

**IT'S LIKE PHOTOGRAPHY,** the golden hour, when I want the best picture of this mountain when the shadows are long and the light is changing from the bright of day to night.

.......................................

**THE TERMINATOR, THAT'S** where you can really see where the moon is a world, a rocky world.

.......................................

**ONCE, I WAS** lost, I drove past Tuscon because I'm a terrible navigator and there was lots of construction. I called this guy I had gone on one date with, because I was freaked out, all the lights had disappeared and I didn't know where I was or how to get back. He asked me where I was and where I was going, and I said I didn't know. "Where is the moon?" he asked. I told him. "Well, if you get it into the back passenger window, you'll get back." And he was right. He got me home by the moon. And that guy? He's my husband now.

.......................................

**THE MOON IS** still my favorite thing to look at, nothing to me is as beautiful. It's so real, so big, and from there it's just a stepping stone to the rest of the sky.

# CHARLIE DUKE

*ASTRONAUT*

**WHEN YOU GET** out the first time, you are in wonder, awe.

......................................

**FIRST, THERE'S THE** beauty of the moon, the brightness of the surface in the sunlight, the intensity of the shadows. The moon itself was mostly shades of grey, though some rocks were black and glassy-looking. And some were white, these were rocks that were shocked by a meteorite impact.

......................................

**THE ONLY COLOR** at all was the Earth, all blue from the oceans and white from the polar icecaps and clouds in between. From where we stood, it was directly overhead.

**ON THE MOON,** you had this feeling of belonging. I felt right at home the whole time.

......................................

**YOU RECOGNIZE THE** landmarks that you've spent a year studying. You're in constant communication with Houston. Plus, [Appollo 16 Commander] John Young and I were always talking. I mean, you can hear one another breathing. Every breath triggers the transmitter. So it never felt like we were in a hostile environment.

......................................

**THE MOON WAS** just an exciting, intriguing place. What is over the next ridge? It's just the feeling of adventure. Let's go explore.

**THE BIG MOUNTAINS** were to the north. They were probably 500 to 600 feet high, but so far away that we couldn't reach them. Stone Mountain was to the south and about four miles away. It was a big, smooth, rounded-top hill that went 300 or 400 feet above the valley floor. It looked like the Stone Mountain outside of Atlanta.

**ON THE SECOND** day we drove almost to the top of it in the lunar rover, to an objective called Cinco Craters. There were five craters that they wanted us to investigate.

**DRIVING THE ROVER** was rough. The suspension was designed for the moon, but when you hit these bumps and little craters at 1/6 gravity, or ran over rocks, it bounced. A lot.

**GO FIND THAT** Lunar Rover Grand Prix footage I took of John driving. His wheels were off the ground in some instances. When I rode with him it felt squirrely, almost like driving on ice in the wintertime. You fishtail a lot.

**IN TRAINING, JOHN** and I were always joking around, so we decided that's how we'd behave up on the moon. It was an Olympic year, 1972, so we decided to have the Lunar Olympics.

**WE WERE GOING** to go for the moon high jump record. But when I went up, my center of gravity went backwards. That backpack weighed as much as I did, so normally you'd bend forward to keep your balance, to keep the center of gravity through your feet. But when I jumped I straightened up, and that brought me over backwards. Next thing I know, I'm flat on my back.

**THERE'S THE EARTH,** right out the front of my visor—but I wasn't focusing on that at all. I'm thinking that if my backpack doesn't hold, this was going to kill me. Thankfully, I survived. John made a comment, "That wasn't very smart, Charlie." I said, "Yeah, help me up."

**THE SUIT HAD** some restrictions. You can't run in it like you can in a pair of shorts. Running across the flat moon was what I'd call a stiff-legged jog. Going up a hill was more of a hop, and going down was a skip.

**BUT ONCE YOU** got going, you really felt free.

**I'M GONE** a lot. My family—my wife and two sons, one just turning seven, and one almost five—

lived in Houston, but we trained mostly in Florida. One day I asked the boys, "You guys want to go to the moon with dad?" Of course they said yes.

. . . . . . . . . . . . . . . . . . . . . . . . . . . . . . . . . .

**A FRIEND FROM** the NASA photo shop came over and took our picture. And then I got permission from Deke Slayton to take it and leave it on the moon. If you look at that picture up close, you can see everybody smiling like crazy.

. . . . . . . . . . . . . . . . . . . . . . . . . . . . . . . . . .

**ON THE BACK** of the photograph, I wrote, "This is the family of Astronaut Charlie Duke, from Planet Earth, who landed on the moon on the 20th of April, 1972."

. . . . . . . . . . . . . . . . . . . . . . . . . . . . . . . . . .

**THE DAY AFTER** launch, [Command Module Pilot] Ken Mattingly misplaced his wedding band. He put it in a pocket, but when he went to retrieve it, it had floated out and disappeared into the spacecraft.

. . . . . . . . . . . . . . . . . . . . . . . . . . . . . . . . . .

**SO ON OUR** way to the moon, we're looking everywhere for his wedding band. Then John and I got in the lunar module. We go, land it, come back three days later, and Mattingly still hasn't found that ring. It has to be in here somewhere, but we can't find it.

**ON THE WAY** back, we had a spacewalk. We're moving along, and we open the hatch, and Mattingly gets out and goes to the back of the Service module to do some experiments.

. . . . . . . . . . . . . . . . . . . . . . . . . . . . . . . . . .

**I LOOK OVER,** and there's this wedding ring floating out the door.

. . . . . . . . . . . . . . . . . . . . . . . . . . . . . . . . . .

**I REACH FOR** it, but I'm wedged in down there and couldn't catch it. And it floated right out the hatch.

. . . . . . . . . . . . . . . . . . . . . . . . . . . . . . . . . .

**IT WAS MOVING** so slowly. The only relative motion was just a few molecules of air that were still escaping from the spacecraft, and they carried the ring out towards deep space.

. . . . . . . . . . . . . . . . . . . . . . . . . . . . . . . . . .

**IT FLOATS UP** towards Mattingly, who's got his back towards me, and I'm thinking it's going to hit him. But he was busy on the experiment so I didn't say anything. And then it did hit him—a round ring on the back of a round helmet— and it took a perfect, 180-degree bounce and started back towards the hatch.

. . . . . . . . . . . . . . . . . . . . . . . . . . . . . . . . . .

**MAYBE A MINUTE** or two later, it floated right back in and I grabbed it.

Charlie Duke family portrait left on the moon. Photo credit: Charlie Duke.

# MARK BENDER

*ECLIPSE CHASER*

**ECLIPSE CHASERS ARE** crazy people that at some point saw one and thought, "Oh my God, what was that? I have to see that again."

**WE GET CALLED** umbraphiles, because the moon's shadow is called an umbra.

**I THINK** it's a misnomer, since what we're actually observing is the corona. Coronaphiles would be more like it.

**DURING AN ECLIPSE,** you can see the sun's corona, or crown, which you can't with the eye under any other circumstances. It's these white light emanations, made up of ionized particles coursing out in long, white, willowy streamers that seem to have structure.

**OBSERVING** it is a really rarefied moment for human experience.

**ECLIPSES HAPPEN** because the moon is 400 times smaller in radius than the sun, but the sun is 400 times further away. So they appear to be the same size in the sky from Earth, which is ridiculously interesting. Quite a coincidence, right?

**I AM A SPIRITUAL** person, and in my book, that's intelligent design.

**THEY'RE NOT** *exactly* the same size. If they were that perfect, an eclipse would last an instant. But they're a little bit out of whack, and that gives you the three or four minutes, up to the maximum total solar eclipse of seven minutes. So it's kind of an imperfect perfection.

**MY FIRST ECLIPSE** was in Cornwall, England, on August the 11th 1999, right on the verge of the year 2000, the new millennium, apocalyptic change.

**IT WAS** actually at 11:11 on the 11th, which I thought was interesting.

**CORNWALL WAS** a brilliant place, but the odds were 95% that we wouldn't see it because of bad English weather. Then, right as it started, the clouds parted like the Red Sea, and the eclipse presented itself.

**I SAW PERFECT** alignment of the sun, the earth, the moon and

me—the human being. And I witnessed the corona. At that moment, I said to myself, "Wow, this is embarrassing. Yesterday, or even 10 seconds ago, I was an atheist-scientist-professor at a university, and now I totally believe in God."

**I DIDN'T THINK** about why. I didn't think, "Wow, this is so beautiful and overwhelming... it must be divine." I actually didn't think anything. I just suddenly believed in God, out of nowhere.

**THE MOON LOOKS** enormous up there, like this this big, giant, jet-black dot. And then you get this ephemerous, white, milky-looking streamery pushing off in all directions from it. Then before you can even take a breath, the experience is over.

**THE LONGEST ONE** I've ever seen was four minutes and 40 seconds.

**THE MOON'S EDGE** is called the lunar limb. And the moon's got mountains all over it, so the limb is bumpy. Depending on where you are in an eclipse's path, you'll see sunlight dip between the peaks, producing these ripples of gleaming light called Baily's beads.

**WHEN IT** goes dark, the temperature will drop up to eight degrees almost instantly.

..................................................

**ANIMALS WILL** think it's night and go into their night behavior. I saw an eclipse in West Texas and a bunch of wolves started to howl.

..................................................

**IN 2015,** I saw one on this small island called Spitsbergen, in the Svalbard archipelago, which is the last land mass before you get to the ice sheets of the North Pole, near Norway. It's full of polar bears and young people who carry rifles. It was twenty degrees below zero.

..................................................

**THAT WAS** the first time I saw shadow bands.

..................................................

**THEY'RE THESE** little, rippling waves of shadow that undulate on the ground, like if you shine a flashlight into a swimming pool.

..................................................

**THE MIND-NUMBING** part is that it takes about 18 months for another one, and you've usually got to go to the other side of the planet.

..................................................

**AND THEN YOU GET** another couple of minutes, at the end of which you think, "Oh, my God, that wasn't enough. I have to see that again."

..................................................

**SO IN MY CASE,** you end up running around the planet for 20 years, in which time I've got about 18 minutes of experience of the corona of the sun.

..................................................

**IT'S NOT** enough.

..................................................

**THE MOON IS** moving away from the earth. We're talking a quarter of an inch every year, millimeters. At some point it will be so far away that it will not appear the same size as the sun, so when they cross paths, the moon won't block out all the light.

..................................................

**THAT WILL BE** the end of solar eclipses. So for now we're in a period of grace. That will end in a half-million years.

Lunar Eclipse on Easter Island. Photo credit: Mark Bender.

# CHRISTINE DARDEN

*AEROSPACE ENGINEER*

**I GREW UP** in Monroe, North Carolina, the youngest of five children. I was eight years younger than the next one, so pretty far behind.

.....................................................

**BY THE TIME** I was eight, they'd left home, and I was almost an only child.

.....................................................

**WE GREW UP** in an entirely black neighborhood. I did not come into much contact with whites.

.....................................................

**I HELPED MY** dad work on the car, helped him in the yard, went to work with him even. I liked handling things. I could fix the brakes on my bike with a coat hanger.

**MY MOTHER TAUGHT** school, out in the county, and she took me with her when I was three. I started learning early. By the time I went to school in Monroe, I was two years ahead.

.....................................................

**SHE GAVE ME** a talking doll once and was disappointed that I didn't play with it much. I cut it open to see why she talked.

.....................................................

**I FELL IN** love with geometry at an all girls boarding school in Asheville. It was the problems that fascinated me.

.....................................................

**SPUTNIK** went up my senior year.

**I REMEMBER** October 5, 1957. The headline of the paper was that Sputnik had been launched by the Russians, flying 17,000 miles an hour around the Earth. Because I'd grown up doing the drills, hiding under our desks and in safe rooms, the news was quite significant.

. . . . . . . . . . . . . . . . . . . . . . . . . . . . . . . . . . . . . .

**I WENT OFF** to college in 1958 and decided I wanted to major in mathematics. But my dad thought getting a teaching certificate would be wise. So for every elective, I took the higher level classes all the math majors were taking.

. . . . . . . . . . . . . . . . . . . . . . . . . . . . . . . . . . . . . .

**I WENT TO** the placement office and told the young lady I was getting ready to graduate and looking for a job. She said, "Didn't you know NASA was here yesterday recruiting?" She had me fill out an application and she mailed it to them.

. . . . . . . . . . . . . . . . . . . . . . . . . . . . . . . . . . . . . .

**THREE WEEKS LATER** I had an offer from NASA. That was June 1967.

. . . . . . . . . . . . . . . . . . . . . . . . . . . . . . . . . . . . . .

**I WAS ASSIGNED** to the Hypersonic Complex, where we had wind tunnels to do Mach 22 speed studies, simulating a vehicle coming back into Earth's atmosphere from space.

. . . . . . . . . . . . . . . . . . . . . . . . . . . . . . . . . . . . . .

**HOW MUCH FUEL** would have to be carried to get to the moon? How heavy would it be? What difference would that make in the flight? When they came back, how fast would they come? What angle did they need to be? What temperature?

. . . . . . . . . . . . . . . . . . . . . . . . . . . . . . . . . . . . . .

**WE WERE ALL** extremely excited to sit down that Sunday and turn on the television to watch the Apollo 11 landing. The picture was very snowy, as I remember.

. . . . . . . . . . . . . . . . . . . . . . . . . . . . . . . . . . . . . .

**I WENT TO** the director and I said, "I'd just like to ask you why a male and a female with the same background are assigned to such different jobs."

. . . . . . . . . . . . . . . . . . . . . . . . . . . . . . . . . . . . . .

**HE SAID,** "Nobody's ever asked that question before."

. . . . . . . . . . . . . . . . . . . . . . . . . . . . . . . . . . . . . .

**MY TITLE CHANGED** from data analyst to aerospace engineer.

. . . . . . . . . . . . . . . . . . . . . . . . . . . . . . . . . . . . . .

**"GIRLS DON'T DO MATH,** girls don't do engineering." People have always been telling us that.

. . . . . . . . . . . . . . . . . . . . . . . . . . . . . . . . . . . . . .

**I WOULD LOVE** to see vehicles that can take off and go into space maneuver more like airplanes. But NASA is pulling everything back. The financing stopped.

. . . . . . . . . . . . . . . . . . . . . . . . . . . . . . . . . . . . . .

**WHEN I THINK** about young girls today, I'd tell them: Go for it. Prepare, work your plan, persist. If this is where you want to go, it will be marvelous.

# KARL MARLANTES

*AUTHOR, VIETNAM VETERAN*

**WHEN I WAS** 14, my cousin and I built a telescope. We were in a logging town on the Oregon coast, with not much to do. But we were pretty bright kids, so we went down in his basement and made a telescope to look at the moon.

**WE SPENT MONTHS** grinding this nine-inch piece of glass into a mirror, polishing it with rouge and pastes and curving it.

**I REMEMBER** the first time we set it up in my backyard. It was a beautiful, clear Pacific North-west coast evening. We took our map of the moon out, turned our new telescope to the moon and just stared. Craters! It was thrilling.

**YOU WANT TO KNOW** where I was when they landed? I was a Marine infantry platoon commander and had been in Vietnam since the fall of '68.

**I'D GOTTEN TWO** purple hearts, so they'd transferred me out of the infantry to the air wing. Put me in the backseat of a little plane, either an OV-10 or an O-1

Charlie, as an artillery spotter. Pilot in front, me in the back.

**THEY CALLED US** forward observers.

**THAT DAY** we had a mission to fly. Nothing was happening. I can't even remember the mission. It might have been protecting a convoy, flying ahead to see if we could spot ambushes. Something like that.

**I REMEMBER IT** was about 10:00 in the morning, and we realized the moon landing was going on. Well, I'm sorry to say it, but we just took the airplane as high as we could. I don't know how high that was. We just kept going. Our only limit was oxygen.

**WE PICKED UP** a radio station in English, either from Singapore or Australia. And we're way up there, as high as we can go and not get apoxia, listening to the moon landing. It was unbelievable.

**I'D BE AMAZED** if the troops on the ground knew about the moon landing when it happened. I'm sure at some point, somebody got a letter from home or somebody came back from R&R and said, "Hey, they landed on the moon!" It

wouldn't have taken long for word to get around after that.

**IT WAS ONLY** good luck that I happened to be in an airplane and the pilot was interested as well.

**MOONLIGHT** is extremely important in combat operations. If you're on an ambush, seeing the moon probably means there's too much light—or just the opposite, because if you need to see the enemy, you need that light.

**YOU'D PLAN THINGS** based on a full moon or a new moon or the time of the moonrise. So the moon determined tactics. At least it used to.

**UP UNTIL THAT POINT**, America was kind of wobbling. I had just been in Europe, studying at Oxford, and we actually sewed Canadian flags on our backpacks so we could get picked up when we went hitchhiking. People hated Americans.

**SO TO GO** from that, and then be fighting in this war...I mean, I was very proud to be a Marine. But to be an American? Not so proud. Then we land on the moon, and it was like, Goddamn!

# ED DWIGHT

*SCULPTOR, ASTRONAUT CANDIDATE*

**MY MOM WAS** ahead of her time. She spent a lot of time teaching us about self-confidence and how to not allow segregation to dictate our lives.

**SHE WAS** Catholic, and she went to Mass every day.

**WE LIVED** on a small farm outside of town, and it butted up against Fairfax Airport. We'd walk down to the airport, and when it was dark, mom would teach us about astronomy and the moon and orbital mechanics and the Milky Way and the sun.

**WHEN I WENT** into space training, I remembered it all. How she knew this stuff, I can't even.

**WE WERE** the first black kids in the only Catholic school in the city. This was 1947, before Brown v. Board of Education. My mom kept writing the Vatican until they ordered the school to integrate.

**THE DAY I WALKED** in the door of Bishop Ward High, over 300 students pulled out and went to the public school.

**I HAD A JOB** at the airport cleaning out the planes. They'd give me a nickel or a dime for a plane.

**WHEN THEY** turned Fairfax into a training center for fighter pilots and they put up a 20 foot wall, I started to draw everything in the Air Force inventory. And I went to the library and I read all seven Air Corps manuals.

. . . . . . . . . . . . . . . . . . . . . . . . . . . . . . . . . . . . . .

**I TOLD MY DAD** that I wanted to be an artist, and he went ballistic. "No you're not," he said. "You're going to engineering school." I told him, "I don't want to drive any trains!"

. . . . . . . . . . . . . . . . . . . . . . . . . . . . . . . . . . . . . .

**WHEN THE** Air Force Academy came to my school, I was the first one in line. Because I stuttered, I didn't say a word. But I passed everything. And they sent me out to Denver to take more tests. When I opened the booklet, sure enough, it was the exact same test as the one in those flight manuals from home.

. . . . . . . . . . . . . . . . . . . . . . . . . . . . . . . . . . . . . .

**THEY CALLED** my mom and told her I was a genius. I went right into the pilot training program, and ended up at the top of my class.

. . . . . . . . . . . . . . . . . . . . . . . . . . . . . . . . . . . . . .

**I WAS BASED** at Travis Air Force Base, north of San Francisco, when I got the letter.

. . . . . . . . . . . . . . . . . . . . . . . . . . . . . . . . . . . . . .

**THREE YEARS AFTER** NASA was formed, President Kennedy came up with the idea to make a black astronaut.

**THE GUYS** in my unit told me, "Don't do it. You're safe. You're set. You're going to have a star on your shoulder."

. . . . . . . . . . . . . . . . . . . . . . . . . . . . . . . . . . . . . .

**I SENT MY STUFF** in and four days later, I got orders to go interview with Chuck Yeager.

. . . . . . . . . . . . . . . . . . . . . . . . . . . . . . . . . . . . . .

**IT WAS FOUR GUYS**, all with stars and eagles on their collars, beating me with questions, asking me, "How would you feel on the end of a rocket?" And at the end of the deal, the last question was "Are you happily married?"

. . . . . . . . . . . . . . . . . . . . . . . . . . . . . . . . . . . . . .

**THEY HAD** all my paperwork sitting in front of them. I told them, "You must know that I am divorced."

. . . . . . . . . . . . . . . . . . . . . . . . . . . . . . . . . . . . . .

**AND THEY SAID,** "Go find her." They needed everything to look normal, a story for the public. So I found her and we got back together.

. . . . . . . . . . . . . . . . . . . . . . . . . . . . . . . . . . . . . .

**THE PRESIDENT** called my parents the night before the announcement to congratulate them. He told them that it was good that we were from Kansas, because Kansas was a free state.

. . . . . . . . . . . . . . . . . . . . . . . . . . . . . . . . . . . . . .

**AND FROM THE** very beginning, after he announced me as the first black astronaut candidate, it was doomed.

**SENATOR** Strom Thurmond told the President, "If those people see a black man in space, they will not understand why they can't vote."

**WHEN I WAS** at the Manned Orbital Laboratory, as the thing went on, I realized, "Damn, I can do this."

**I WAS ON** magazine covers all over the world. Jet, Ebony, Sepia. I was all over the news. Making speeches all over the country.

**CHUCK YEAGER** would call me into his office and tell me that I was only there to get Kennedy the black vote. He told fellow students not to fly with me. He told me, "Why don't you quit now, because you know you'll be dead."

**WE WERE ALL** in Seattle, at a Boeing event for their space simulator. It was maybe 10 o'clock in the morning, and Kenny Weir comes in and tells us the that the President had been shot. He thought the Vice President had been, too.

**MONDAY** morning, back at Edwards Air Force Base, I had orders in my box, shipping me to Germany. That's how fast it took them to get rid of me.

**I JUMPED** in my plane and flew to Washington and went straight to Bobby Kennedy's office. And he killed those orders. Then the fight was on for what to do with me.

**THE EDITORS** at Ebony ran a story that had me calling President Johnson a racist. That blew everything up.

**I PUT ALL** my awards in my little Volkswagon and just drove off the base. Goodbye Air Force career. I cried all the way to Denver.

**I ALWAYS HAD** a fascination with the moon. It was almost a kinship. Going all the way back to those walks with my mom. She'd make up stories about every kind of moon—crescent, half, full. I'd be gawking at the moon, it would be close enough to touch, and she'd pull these incredible stories out of her bag. She'd conjure up these things and our mouths would be wide open.

**I'D ASK HER,** "Can you go there? How long does it take? Has anybody ever been?"

Captain Ed Dwight, U.S. Air Force.
Edwards Air Force Base in California, 1963.

# SUSAN MILLER

*ASTROLOGER*

**PEOPLE ASK ME,** "How do you know what's going on in the heavens?"

**NASA PUBLISHES** the calibrations. There is such a thing as an ephemeris. That is a table of planets. Google it, and you can see where the planets are for each day.

**ASTROLOGY IS** the study of cycles of the planets, some of which will repeat, and some of which will never repeat, in our lifetimes. The planets are just too far apart. And the space between the planets, how they align, is how we interpret things.

**THE MOON COMES** at night, you don't see it as often.

**THE MOON IS** the fine-tuning to your character. It is the part of you that people get to see when they get to know you well. It is the repository of your memories and dreams.

**AND ALSO** the moon is how you viewed your mother.

**LET'S SAY** there are five children and each has the natal moon in a different place, they would each describe their mother differently. Like the blind men and the elephant. They all see a difference.

**THE DAY OF** the first moon landing was such a glorious chart that I wonder if NASA had an astrologer help pick the date. But I guess not. You know scientists.

..................................................

**YOU RARELY** see a chart this positive.

..................................................

**PISCES WAS** on the horizon line, which is spirituality and helping other people. That's what Pisces is all about. It's a spiritual, religious sign even.

..................................................

**WHAT I FIND** fabulous is that the moon was in Libra, conjunct Jupiter! Oh, my God! Jupiter's the giver of gifts and luck, and the moon is publicity. So everybody was excited, happy...and Uranus was there too!

..................................................

**ALL THREE WERE** on the head of a pin. You had the moon, Jupiter and Uranus all lined up.

..................................................

**URANUS IS** in charge of creativity. It rules high technology, any of the sciences. But it rules the humanities as well. When we try to help people like battered women or the environment, that's all under Uranus trying to help the world be a better place.

..................................................

**URANUS ALSO RULES** groups of people. So I guess the United States was the group, but because

Uranus was conjunct Jupiter, the giver of gifts and luck, we were giving something to the world that day.

..................................................

**JUPITER AND URANUS** were conjunct in the house of partnership. That's the breakthrough aspect. This was a tremendous team effort.

..................................................

**IMAGINE BEING** a teacher to these eight little planets. Now, the sun and the moon don't retrograde but the planets do. And you're a teacher trying to get them all together. Uranus, stop jumping around and throwing paper airplanes. Saturn, stop sulking in the corner, get over here. Venus stop kissing Mars, sit down, come on.

..................................................

**I MEAN,** they're a bunch of little rebels. It's very hard to get them all together. And on that day, not one planet was retrograde. [Gasps] This was such a good day, it was insane!

..................................................

**YOU HAVE WHAT** we call a Grand Trine to the Ascendant. Have you ever heard people say, "What's your rising sign?" The "rising sign" is the sign on the Eastern horizon as you're born. It creates this golden triangle with the sun and Mercury, real close buddies that stay together—on this day one is 26 degrees, one's 28—and they're sending beams up to Mars and Neptune. So that's a trine, 120 degrees. It's as heavenly as you're ever going to get.

**ON ONE SIDE,** the sun and Mercury are in Cancer, because it's July, and they're sending these beautiful kisses and beams up to Pisces rising. And you've got Neptune, which is Pisces's ruler, right next to Mars, which…Neptune is inspiration, sometimes confusion. It was in the money house, so they might have had a little problem with the budget.

**WHEN MARS** is good your actions lead to success. When Mars is not so good, you run into obstacles and have to fix them. But this was so heavenly. Not only did you have this Grand Trine, you have Pluto in the house of work, at a perfect, 60-degree angle—which means opportunity—to the sun and Mercury, in the house of creativity.

**MARS WAS SO** friendly. He was in Sagittarius, which rules global interests. It's the sign of the international traveler. This is crazy! This is beyond international!

**SATURN WAS** on the moon. They sweated through this. This was hard. They were so relieved when they got there. Saturn is the lessons we learn in life. It's the hard work that we do, but that we are proudest of in the end.

**URANUS WAS** talking to the sun. When the sun and Uranus speak together, that is the sign of breakthroughs.

**THAT IS WHY** I think somebody at NASA must have known astrology.

**I REMEMBER** the whole family watched—my sister, my mother and father. And we turned on the TV too early, just to be sure we wouldn't miss anything, to see if the astronauts were okay.

**I WAS JUST** a kid in high school, and I didn't know what the moon meant. My mother knew astrology, but I didn't. And I remember getting tired but being determined to stay glued to the TV. It was like two or three in the morning when it finally ended.

**DID THE MOON** landing have an effect astrologically? Not that we know of.

Creti, Donato. *Astronomical Observations: The Moon.*
1711, Pinacoteca Vaticana.

# STEVE BALES

*APOLLO 11 GUIDANCE OFFICER*

**AS A BOY** in Iowa, on hot nights, my dad, my brothers and me would go out and look at the sky. The stars were magnificent, like they were coming down on top of you.

**DAD KNEW** a bit about the constellations, the Big Dipper, the North Star, Orion.

**OCTOBER 1957,** Sputnik happened. You can't imagine the anger, anxiety and frustration of the whole country.

**I MADE UP** my mind to go into aerospace engineering.

**I CAME ON** at the end of Gemini 3, and the pressure was on to get a man on the moon by 1969.

**FOR APOLLO 11,** I don't think anybody in the decision-making positions at Mission Control had anything other than a bachelor's degree. Most of us were the first in our families to go to college.

**THE PERSON** responsible for the propulsion tanks and engines, how they would be armed, how they came on, that guy, Bob Carlton, was actually a high school dropout.

**BUT GENE KRANZ,** the flight director, said something once that stuck with me. "There are poets, and there are plumbers. The poets write the equations and the software. They do the theoretical stuff. We're the plumbers."

..............................................

**YOU'D BE** amazed how much physics a plumber understands, too. They know darn well that if the pipe is too small, that's going to be a problem. Or if it's too big. And that's us.

..............................................

**MY JOB WAS** the computer and instruments needed to navigate in space.

..............................................

**WHEN THE ENGINE'S** not burning, Newton pretty much knows how things work—Newton, plus a lot of smart guys to tweak him up. Then you're just floating, controlled by the gravity of different things. When that engine is burning, you control your own gravity.

..............................................

**YOU'RE TRYING** to pinpoint-land a couple of guys in a spaceship, and you have to start that process at the right altitude and the right speed. And the moon, it turns out, is much less uniform than the earth.

..............................................

**THAT SUNDAY** morning we came in at about 7:00 a.m.

**WE'RE SENDING** information up to them, telling them where they are, where they will be when they land, what time they should start the engine and a number of other, critical check-out things.

..............................................

**FINALLY, EVERYTHING IS SET:** the computer is fine, the propulsion, communications—everything looks good.

..............................................

**THEY WILL GO** behind the moon one last time and lower to 50,000 feet. The next time we'll see them, we'll have 15 minutes to look at everything. Then we'll start the engine, and it'll burn for 12 minutes from start to landing.

..............................................

**IN ANTICIPATION** of that, we take a break. Gene says, "Everybody be back here in 15 minutes."

..............................................

**WE ALL** come back, and Gene says, "Everybody get on the Special Loop." We had these communications loops. Whenever you punched one up, you talked to different people. A couple loops were really important. One was the Flight Director Loop. Punch that up, and you're talking directly to Gene, so you didn't use that one unless you had to.

..............................................

**IF YOU'VE EVER** listened to old Apollo 11 recordings, that was the Flight Director Loop. All the loops went different places. The Flight Director Loop was relayed to the public as was the Air-to-Ground loop—NASA, CBS, NBC, you name it. But the Special Loop was private. It went nowhere except inside the control room.

**GENE GETS ON,** and here we are. Just think about it. We are 10 minutes away, all 26 or 27 years old, and we know that if something goes wrong in our area, that might stop the mission, or worse.

**HE SAYS,** "Okay, we have trained for this, and we're ready. All our lives, we've worked for this moment. But I want you guys to know something clearly. When we walk out of this room, no matter what happens, we walk out as a team." He didn't have to say anything else.

**THERE WOULD** be no finger pointing. We walk out together, good, bad or indifferent.

**IT LOOKS PRETTY GOOD** until just as they start the engine, you hear nothing but noise. But we're able to yaw the ship a little bit left to right. And then I get a problem. The computer doesn't realize we're going toward the moon about 14 miles-an-hour faster than we should be. If that had been another seven or eight miles-an-hour faster, I would've had to call an abort.

**I TELL GENE,** and right away he knows we're close to a big problem. But it might not grow. About a minute later I say, "This error isn't growing. It looks like we're going to be okay." In fact, I added the two words I never should say. "We're going to be okay, I think."

**THEN, THE CREW** reports a computer alarm. So right when I'm thinking my problems are over, they're only starting. This alarm is never supposed to come up.

**IT'S A** four-number readout, 1202. That's all the crew has. They don't have fancy displays like we have today. They don't even have a fuel gauge. They just have numbers. And this number comes up. Twelve-Oh-Two.

**THANK HEAVENS,** we had gotten a similar alarm in a simulation, so we have a rule on how to react. And I've got the software expert yelling it in my ear, "We're okay if it doesn't come too fast!"

**AS WE KEEP** getting closer, alarms keep coming.

.................................................

**INSTEAD OF** coming into this nice, smooth area, Eagle is going down somewhere we don't quite recognize. Almost into a big crater.

.................................................

**THEY'RE STILL** getting these alarms. One happens so bad that Neil's [Armstrong] whole display goes blank for 10 seconds, and he's trying to figure out where to go. Finally he just makes up his mind—he goes into semi-automatic program where he controls the vehicle's direction and the computer controls the altitude rate. This lowers the computer's work load and the alarms quit, thank the Lord, at about 750 feet.

.................................................

**BUT HE'S FLYING** all over the place. The fuel sensor says we just hit low. We've got 60 seconds of fuel left.

.................................................

**SO BOB CARLTON,** the propulsion engineer, starts his stopwatch. I'm telling you, the whole lunar landing is counting on a man with steely nerves and a stopwatch.

.................................................

**AND IF YOU** listen to old recordings, you hear "60 seconds!" That's Bob calling to the flight director. Then "30 seconds!" If he goes to zero, we gotta leave. But Neil finally sees a place that's good enough, and he brings it down, and we land.

.................................................

**ARMSTRONG SAYS,** "Tranquility Base here," and that was a new name for us. They'd always called themselves Eagle. And I just thought, "Hey, that's a great name." Then we all went back to work.

.................................................

**MEN AND WOMEN** have looked up at the sky for centuries. And suddenly, wow. There's somebody on that.

.................................................

**I READ** something a couple of years ago that said in the year 3000, if there is a name from the 20th century that is widely known, it would probably be Neil Armstrong. I think that's right.

# DAN WINTERS

*PHOTOGRAPHER*

**WHEN I WAS** a kid, I wanted to be an astronaut. Most of my friends did as well. The real heroes were coming out of that world.

**IT WAS SUMMER,** so I was out of school. My mom and dad were really good with fostering my interests, so they let me watch every minute of Cronkite, sitting in my dad's chair, building Apollo models on those TV trays.

**I'VE JUST FINISHED** a six-month project on the moon landing for *National Geographic.* I shot thousands of photographs, everything from its effect on popular culture—like Apollo lunch boxes—to the Apollo 11 capsule itself. NASA gave me six hours with it.

**I SHOT NEIL** Armstrong's gloves. Gene Cernan's uniform. Jim Lovell's suit. Pad 39A, the launch complex Apollo 11 lifted off from.

**LOOKING AT THE** Apollo 11 capsule brought home just how tenuous those missions were. A Saturn V rocket was made up of over a million pieces. Well over. And the acceptable failure rate is pretty much zero.

**FOR THIS THING** to work, everything has to operate to 100%.

IT IS AMAZING to be in the presence of these artifacts, to have these full-circle moments I never could have imagined as a boy.

.............................................

I LOOK AT Apollo as the greatest, most expensive photographic platform in history. Because images are what connect us to that expedition.

.............................................

NASA RETAINED Hasselblad to make special cameras and to come to the States to train the astronauts in photography.

.............................................

THEY LEARNED IT the same way they learned geology, by doing. That's what I loved about the Apollo program.

.............................................

THE CAMERAS ON the moon didn't have viewfinders, so the astronauts had to learn how far away from a subject they needed to be to photograph it. The camera itself was mounted on their body. It had three apertures for side-lit, back-lit or front-lit; an electric film advance, so all they had to do is push the button; and a gigantic, 250-exposure magazine.

.............................................

ARMSTRONG WAS THE designated photographer on the moon. There is only one photograph of him on the lunar surface. He can be seen as a reflection in Buzz Aldrin's visor.

THE CAMERAS THEY took to the surface? They discarded those. Just threw them out. The astronauts took the film-backs off and tossed the cameras out of the LEM before they closed the hatch.

.............................................

THINK ABOUT ALAN Bean [Apollo 12] and what he brought back to his painting and artistic pursuits. These astronauts were all engineers who became geologists and photographers and artists and poets.

.............................................

BILL ANDERS [Apollo 16] shot Earthrise as they came around from the dark side of the moon and the Earth revealed itself.

.............................................

AND HERE'S WHAT'S interesting about Earthrise: It's difficult to establish orientation in space because there's no actual horizon anywhere, right? But the way Anders originally shot it, the moon was on the right side of the frame and the earth was on the left. When the film was being processed, the guys in the NASA film lab passed over it. It was just another Earth photo for them.

.............................................

BUT IT'S THE relationship between the moon and Earth that would make it powerful.

**WHEN SOMEBODY** turned that photo on its side, and the moon was the horizon and the earth was reoriented to the top of the frame...that's when it gained traction. It looked like the earth was rising, and they gave it that title. That cemented the idea that, if there's another inhabited celestial body, this is what we'd look like to them.

**FOR ME, THE** more important image was of the full Earth shot by Apollo 17. Either Gene Cernan or Harrison Schmitt took it—nobody really remembers because they were all shooting.

**I DON'T THINK** they realized how significant that image was. But that was the first time we'd seen the full earth, the first time it had been shot with the sun at the photographer's back so that the entire globe was illuminated. To this day, it's the only full-sphere image that exists, the only one that shows us where we live in its entirety.

**I THINK THEY** shot it from 18,000 miles away. But that was the last Apollo mission. No one has been even close to far enough away since then to take that photograph. There's never been another time where man was able to stand and contemplate his home.

**IN MY OPINION,** it's the most important photograph ever made. Honestly.

**IT'S TOTALLY** in the public domain, and really, its power has almost been erased. People don't realize it's a sacred piece. But if you look at it quietly and contemplate what that image is, it's incredibly powerful.

**THERE'S A GENRE** in photography, a pre-internet genre, of people shooting an image of an historical event as it's being shown on TV. Back then, that was the way to capture it and keep it for yourself.

**MY DAD WOULD** do that. For Apollo 11, he had our camera in front of the TV, and he was really excited. He counted down with the launch—"Three...two...one!"—then started taking photographs. I remember it plain as day.

**THE LAB DATED** the prints, "July 1969," which is funny because we were not that family that took a lot of photographs and rushed to have them developed. But he didn't wait with this. And the really funny thing is that his photographs didn't come out. He was so excited that he left the flash on. So it's a picture of our TV set, and the screen is completely white.

Larry Winters, Moorpark, California. July 16, 1969.

# ANGIE RICHMAN

*NATIONAL PARK RANGER*

**ARCHAEOASTRONOMY,** or cultural astronomy, is the study of past cultures and how they use astronomy in their lives to tell time, to plant and harvest crops, and how a lot of that astronomical knowledge is representative in the material cultures that they left behind—in buildings, in rock art.

**MY BACKGROUND IS** in astronomy. I have a degree in astrophysics.

**THE WAY I** got into the parks service—it actually had nothing to do with the parks service. It was more to do with the sky.

**I KNEW FROM** a young age that I wanted to go into astronomy.

**THE THOUGHT** that people had walked on the moon, that we had been there—it captivated my imagination.

**I TOOK A** semester off school and went down to Chaco Canyon as a volunteer. I was working 40 hours a week doing just the normal ranger job. And then every night—seven nights a week—I worked with the astronomy volunteers.

**THE MOMENT I** saw a picture of Pueblo Bonito, which is one of the largest prehistoric buildings in the park [Chaco Culture National Historical Park], I was like: I have to go there. And we went there and they had a couple of astronomy volunteers set up with telescopes on the patio for us that night. And I just fell in love.

**NATURAL BRIDGES** National Monument has one of the most pristine night skies that I've ever seen.

**IT'S SO REMOTE** and so far away from any city light—it's the closest you can get to a true night sky. The same sky that our ancestors saw a thousand years ago.

**PEOPLE ARE SOMEWHAT** disconnected from the night and from darkness.

**I TEND TO** look for the moon first. And then I'll look for any planets, because they're the next-brightest. And then I'll look for the constellations and kinda walk my way, usually from east to west. And then I'll look north to see where the Big and Little Dipper are, where Cassiopeia is.

**I AM A** huge supporter of the Dobsonian telescopes. They're super-portable. It's just the base and the tube. You don't have to spend a lot of time getting the GPS coordinates right or trying to figure out the GoTo system.

**YOU BASICALLY** just pop it on the base and it's ready to go.

**GET A STAR** chart and start to learn the sky. It's a lot more gratifying to actually learn to find objects rather than just relying on the simplicity of a GoTo telescope to do it for you.

**I DO A** lot of astrotourism. I go to see total solar eclipses, or I travel to places that I know have other prehistoric astrological alignments, like Newgrange in Ireland or to Chichen Itza, down in the Yucatán.

**THAT'S ONE OF** the nice things about the sky: If you can turn off your lights and actually see it, it's just comfort. It's predictable.

**YOU LOOK UP** there and see familiar constellations, kind of familiar faces.

Angie Richman is chief of interpretation and visitor services at Arches and Canyonlands National Parks, Utah.

# TOM JOHNSON

*MEDIA EXECUTIVE & WHITE HOUSE
ASSISTANT TO LBJ*

**PRESIDENT JOHNSON HAD** tremendously close relationships with the leaders of NASA. He'd built friendships with the astronauts. He'd built a tremendous number of allies, Republicans and Democrats. He was a master at that, and President Kennedy knew it.

........................................

**PRESIDENT KENNEDY** appointed [then Vice President] Johnson to head the Space Council, which many people felt was not a particularly major responsibility. But all of us who knew of Kennedy's dream to put a man on the moon within the decade knew that giving it to Johnson was significant. Johnson had been so involved with the space program when he was in Congress, and Kennedy knew that.

........................................

**THIS WASN'T** a toss-away assignment. This was a major commitment of Kennedy's New Frontier. This was the future.

........................................

**IN THE SENATE,** he [Johnson] worked to get as much of the

NASA budget into Texas as he could. Getting that big space center in Houston was a huge triumph.

••••••••••••••••••••••••••••••••••••••••

**JOHNSON'S PASSION** grew out of a combination of things. There was a great worry about the Soviets getting ahead of us in their missile program. There was the tremendous adventure of it. There was the terrific inspiration of President Kennedy's words, all the possible scientific benefits, the spin-off technologies, the spin-off accomplishments. It was astounding.

••••••••••••••••••••••••••••••••••••••••

**ALL OF US** were shaken when Sputnik went up and we saw this Soviet satellite circling the Earth. It injected a good bit of fear.

••••••••••••••••••••••••••••••••••••••••

**THROUGHOUT HIS OWN** presidency, Johnson showed great personal attention to the space program. But I don't think he saw any of that as diverting from his commitment to civil rights and voting rights and the Great Society. I think he saw it as part of that overall idea that this would make our world a better place, that the benefits of the space program would have benefits to programs here on Earth.

**HE GRIEVED** when the three astronauts were incinerated out in that Apollo 1 test. I was on the line with him when he dictated his message of sympathy to the family and the nation.

••••••••••••••••••••••••••••••••••••••••

**HE LEFT OFFICE** anguished that he'd never been able to bring about peace in Vietnam. It was a different world for him. He threw himself into the building of his library, into his book and the televised conversations with Walter Cronkite.

••••••••••••••••••••••••••••••••••••••••

**PRESIDENT NIXON KNEW** of Johnson's interest. They were old political adversaries, but they were also friends. They'd both experienced the trauma of running for and serving in office. So Nixon sent Air Force One to pick up President Johnson and a small group of us in Austin, at Bergstrom Air Force Base, and fly us to Cape Kennedy to see the Apollo 11 launch.

••••••••••••••••••••••••••••••••••••••••

**PRESIDENT JOHNSON HAD** the most premium seat in the house, next to Vice President Agnew. I've got a photo of me standing behind them as we all looked up and watched it soar into space. That engine was enormously powerful. The vibrations were just earth-shaking.

**THERE WAS GREAT** anticipation, terrific hopefulness. But there was a level of concern, too, for the safety of the mission, the safety of the crew. President Johnson knew the astronauts and their families. So there was a sense of relief when it got off successfully, but we still knew it was a long ways from actually getting to the moon and getting home.

......................................

**MY WIFE, EDWINA,** decided to have a moon landing party at our home. A bunch of friends came over, and we tuned in to CBS and Walter Cronkite all the way through it. We were all united with this feeling of absolute awe at what had been accomplished. And to look back now and realize they did it with what we'd today consider rather primitive technology. Just an awesome achievement.

......................................

**NIXON SAID,** "Lyndon, that plaque that we placed on the moon should've had your name on it, not mine." He said, "I've barely gotten into office. You were the one."

---

*Tom Johnson's first role in LBJ's administration was as a White House Fellow, assigned to then Press Secretary Bill Moyers. He held several White House positions, including Special Assistant to the President, and after leaving office, [Tom] Johnson helped lead LBJ's post-presidency ventures in Texas. In 1980, Johnson was named Publisher and Chief Executive Officer of the Los Angeles Times, and in 1990 he joined CNN as President, a job he held for more than a decade.*

Former President Lyndon B. Johnson at the launch of Apollo 11, Kennedy Space Center on July 16, 1969. Photo courtesy of NASA.

# SARAH STEWART

*PLANETARY SCIENTIST*

**THE FASCINATING** thing about the moon is that every culture has come up with an explanation.

**IT'S ON MY LIST** to read mythologies from different cultures explaining the moon.

**EVERYBODY WANTS** the answer to that origin story.

**I WAS A** *Star Trek* fan as a kid. I mean, mainly the old *Star Trek,* with [William] Shatner.

**BEFORE APOLLO,** on the science side, there were major debates of how the moon got to be there. From capturing the moon to having the moon spin off from the Earth.

**APOLLO** definitively showed that all of the preexisting ideas were wrong. Since then, we've known how similar the earth and the moon are, chemically.

**NOW, THE TEXTBOOK** theory is a giant impact. In the early '70s, giant impacts were a radical idea.

**WHEN YOU WORK** out the numbers, the power, the sheer power of a giant impact is equal to the power coming out of the sun every second.

**THE GIANT IMPACT** changes the Earth entirely, introducing an object we've named a "Synestia."

..................................................

**START WITH TWO** Mars-like bodies and collide them. They're temporarily transformed into vapor.

..................................................

**IT WOULD LOOK** like a swirling gas ball with rock dust and droplets cooling at the top of the outer layers.

..................................................

**THE MOON ACTUALLY** grows inside of it. And that is why the moon has this chemical relationship with the Earth.

..................................................

**IN THE LAB,** we test our ideas by understanding the thermodynamics of the materials so that we have some faith in the computer models.

..................................................

**I TOTALLY WISH** we could just collide planets in the lab. That would be so much fun.

..................................................

**WHEN I WAS** a child, I was like, "Of course there's gonna be space travel. And of course people are going to go to all these planets." I had this great hubris of what was possible.

..................................................

**SENDING PEOPLE TO** the moon, bringing lots of rocks back from the moon takes a nationwide commitment. The moon is not ignored.

..................................................

**HOW BIG A** role did the moon have in making Earth, Earth-like?

..................................................

**FOR US** it's understanding where we came from. Understanding the origin of the moon is understanding the origin of Earth.

..................................................

**WHY IS EARTH** different from the other planets? Why are we on this planet and not some other planet?

..................................................

**IT'S A GOAL** to answer that question. And I think it can be answered.

..................................................

**I GO FROM** moments where I think there must be millions of Earths and millions of inhabited planets. But then other days I think, so many things happened in a row to make Earth, and to have advanced lifeforms, will we ever find any? Earth is rare, Earth is hard.

..................................................

**I WOULD REALLY** like us to have a base and use it as part of space exploration.

..................................................

**THE SCIENCE OF** the moon just gets more interesting all the time. It's not a dead planet.

..................................................

**THE MOON** still surprises people.

# STORIES

---

*Personal essays and short writing about the moon*

# ON MOONS

*Written by* **HANIF ABDURRAQIB**

**TO BE FAIR,** I cannot claim that I love the moon as much as all my pals and ancestors and peers. I maybe do not love the moon as much as other poets, who seem to love the moon for what it is capable of doing to the waters, or how it seduces the best or worst out of an astrological sign. I don't know much about astrology, but I do like the idea of astrology for what it brings out in my most creative and magically inclined friends. Elissa, leaning eagerly over a table to ask me if I know the exact hour and minute of my birth, so that we might do my birth chart and finally get down to the issue of what's going on with all my emotional rattling about. Madison, scrolling furiously through her phone over a dinner to see what phase the moon is in, or what planets are twirling ever-maniacally out of whack, so that she might explain to me why all the furniture in my heart's most precious corners has been upturned. Still, I can't say I'm much into what it all means, just that it means something. That we were all born under a different moon and a different sign. And I believe in it, I think! I have taken to waving a dismissive hand and telling a friend "that's such a Virgo thing to say," even when I'm not entirely sure what I mean. And no one has corrected me yet, so either I'm right or I have surrounded myself with immensely kind people, which is probably a very Scorpio thing to say.

Anyway, Robert Hayden loved the moon, and what a fool I would be to not love what Robert Hayden loved. Would be a whole fool not to drink from whatever his palms offered, and Robert Hayden so loved the moon that he decided to strip it of all it's magic so it was just the hanging and cratered glowing rock tasked with dividing up the darkness. Hayden wrote:

Some I love who are dead
were watchers of the moon and knew its lore;
planted seeds, trimmed their hair,

Pierced their ears for gold hoop earrings
as it waxed or waned.
It shines tonight upon their graves.

And burned in the garden of Gethsemane,
its light made holy by the dazzling tears
with which it mingled.

And I, too, dig the moon most when it is a question of its functions. How, for example, I might have once leaned into it in some alley on a clear night to better see the face of a dear brother or to skim a phone number scrawled onto a napkin after spilling out of some dive. I wish to view the moon as Hayden viewed the moon, an object that has a purpose rooted primarily in how it shines, and little more beyond that.

But still, I know that black people in this country have long been obligated to a love for the moon, especially the enslaved who had to traverse the otherwise darkness in a search for freedom, aligning directions with the way the moon fell, and following the shapes of stars. And so I do get the affection for the thing, even if it is sometimes a bit of a showoff, every now and again puffing itself wide and sometimes blushing a gentle red. Dragging me out of the house or interrupting a romantic movie night bending into the potential for more romance, so that we can all go out and stare. Then again, who am I to judge? I've picked through my own closet and opted for the crushed velvet or the bright red on a day where I am dressing for someone else's wedding. I've showed up to the high school reunion in sneakers that have cost more than I made in a week of work at a job slinging books, so I suppose we're all the moon sometimes, depending on the occasion.

While we're here, though, I have to say that I also know nothing of the stars, but have lied about what I know many times. On the television once, a boy traced the freckles of a girl and then pointed at the sky, and she gasped with joy. On a walk in my real life, holding hands with someone somewhere, I pointed up at the stars and pretended to know the shapes of them, and said something about eyes and a promised future, and the person I was with laughed. So okay, I suppose I don't know the stars well enough to lie about them comfortably, but I had a telescope once, bursting out of my top floor window during a time when I lived in a city that got less clogged with a smoky haze during its nighttime hours. And I would look into it every now and then, searching for all the shapes that everyone else saw.

**SO I DO GET THE AFFECTION FOR THE THING, EVEN IF IT IS SOMETIMES A BIT OF A SHOWOFF, PUFFING ITSELF WIDE AND BLUSHING A GENTLE RED.**

But with no luck. From under a campfire, my friend Kyryn said it's easy. The Big Dipper is right above us. And then she traced it out with her finger, but all I saw was a series of tiny explosions that never vanished.

I have to have some sympathy for the moon, for all the foolishness I've projected on to it here, and all it is responsible for when it comes to making sense of the unexplainable human condition. But I also think about how lonely it must be up there among the darkness. Miss Zora, the truest of my ancestors and the only light pouring onto all of my unlit paths, says I feel most colored when I am thrown against a sharp white background which is true sometimes on the train and true sometimes at the birthday party and true sometimes in the office meeting. But what, then, of the sharp and dark background holding up the waning and waxing white, and all of its labor. What I feel is not sympathy, of course. A curious and thankful meditation, but still distance.

---

**HANIF ABDURRAQIB** is a poet, essayist, and cultural critic from the east side of Columbus, Ohio. He is the author of *The Crown Ain't Worth Much, They Can't Kill Us Until They Kill Us* and *Go Ahead in the Rain*.

# WAKE UP

Polar Caves is a roadside tourist trap not far from where I grew up, in New Hampshire. Because we had little money, this kind of thing passed for a holiday trip when I was young. Obviously, the caves are not polar. They are barely caves. They're large boulders with enough space between to allow the passage of parents and their sulky pre-teen offspring. Because I had no car, Polar Caves would be as exotic a destination as I would experience.

Until one cold, clear winter night, when I took a trip by telescope. My high school physics teacher, for whom I quietly, nerdily pined, had let me borrow it. I set it up in the driveway and lowered my face to the eyepiece. In that moment, the moon went from overhead lighting to place. It was a half moon, with crater rims side-lit in sharp detail. I imagined being there among those rocks and peaks, alone, without my parents, possibly with my physics teacher. In that moment, for the first time, I felt the pull of travel. I wanted to explore strange, remote places without gift shops.

It was the first time, also, that I experienced awe. True awe kicks aside the small things in your head and puts in their place something powerful, euphoric, humbling. It makes you say wow when no one is there to hear it. I've chased that feeling, built my career around it, ever since. I've found it in the northern lights and in a total eclipse, in white dunes and black lava beds, in the 3 a.m. blue sky of the south pole. The moon lit that fire. It slapped me in the face and said wake up, little girl, there's a great, mysterious world out there. Go and get it.

---

**MARY ROACH** is an author and science writer in Oakland, California. She has written seven books, including *Packing for Mars: The Curious Science of Life in the Void.*

# I SEE THE MOON

*Written by* **ANNETTE GORDON-REED** | **IF YOU GROW UP** in rural Texas, as I did, in a place without much ambient light, you actually get to see the stars. As a little girl, I took great delight in gazing up, prompted too by a rhyme my mother liked to say whenever we were out at night: "I see the moon, and the moon sees me; God bless the moon, and God bless me."

One evening, I saw what remains the most amazing sight I've ever seen. Stepping outside with my mother after a visit to our neighbor's house, I immediately looked skyward, as I always did, and saw a fiery object moving across the night above us. The stars meant nothing in that moment. What struck me about this startling thing—besides the fact that it existed and had come into my view—was that it made no sound. Something that large, I thought, should have made a sound. The other thing I noticed: There were no visible shooting flames. The fire covering this object looked liquid, like lava from a volcano, except it was up in the sky, in a form that resembled a shucked oyster.

I pointed it out to my mother and the others, who had been saying their goodbyes in the doorway. Their response to this flaming, silent mass was vocal and more intense than my detached puzzlement. "Oh, my God. What is that?" someone cried as we ran into the yard for a better look. We watched as the object sailed silently on before blowing apart far off in the distance. To this day, I don't know what we saw, though I've looked upward many times hoping to see something like it. I never have.

This singular event cemented my interest in the night sky, an interest that drove me, some years after seeing the fireball, to save up and purchase a telescope. It was not a very good telescope, but it was the best I could do with the money I had to spend. The instrument made me feel special, and for a time, "amateur astronomer" was added to my persona of "aspiring writer." I waited for news of meter showers. I looked at—mainly looked at—astronomy books. In the early morning hours, while my family was asleep, I set up the telescope in our backyard and gazed at the stars and the moon—the moon my primary focus because that was what I could see most clearly through my cheap instrument.

Being forty miles north of Houston added to the allure. Houston was "Space City," home to NASA; its baseball team, the Astros, played in the Astrodome—known as the eighth wonder of the world—with AstroWorld, the theme park, right nearby. I grew up with the Apollo program, whose purpose had been declared to the world before I was in nursery school: to land a man [and President Kennedy did say "a man"] on the moon and return "him safely to the Earth." Houston was at the center of this endeavor, and our proximity to the city gave me another strong reason to feel I had come by my interest honestly, as my grandmother would have put it.

As things turned out, I was at my grandmother's home, in Livingston, on July 16, 1969, when a Saturn V rocket launched Neil Armstrong, Buzz Aldrin, and Michael Collins on the Apollo 11 mission. My mother, brothers, and I typically spent several weeks of the summer with my grandparents, but that year, I'd gone alone with my mother. I was feeling a little lonely. And as excited as I was about the history-making voyage to the moon, I was also impatient with the extensive coverage that too severely, I felt, disrupted my television routine.

**THIS SINGULAR EVENT CEMENTED MY INTEREST IN THE NIGHT SKY, AN INTEREST THAT DROVE ME TO SAVE UP AND PURCHASE A TELESCOPE.**

By contrast, my grandparents, both born in the first decade of the twentieth century, were riveted by every moment of the spectacle. My great-grandmother, who watched the coverage with us intermittently, found it difficult to take it all in. I didn't think of it at the time, but she—born in 1880—had come into the world before cars and airplanes existed. And now a rocket was taking men nearly 240,000 miles to an orb she had watched wax and wane for nearly 89 years, likely with no thought that any human would ever reach the place. Her mother was born enslaved but had been freed as a child along with her mother, by her father. From my great-grandmother, to my grandmother, to my mother, to me: I knew people who had known people who had been enslaved in America, and on July 20, 1969, we were sitting together watching live as a man set foot on the moon.

The lunar module landed. Armstrong stepped out and said his famous words, and Aldrin followed. I felt sorry for Collins, orbiting all alone in the command module Columbia, though he did get to see the far side of the moon. Excited and restless after these exhilarating moments, my mother and I decided to take a walk down the road to visit my grandmother's sister and her husband who, if I remember correctly, seemed a bit skeptical of the whole business of moonshots and astronauts. Of course, we looked up at the moon. It appeared the same, but it also seemed different to us, somehow.

As we stared, my mother took my hand. "Just think," she said. "There are people up there."

---

**ANNETTE GORDON-REED** is a professor of law and professor of history at Harvard University. She won a Pulitzer Prize for history and the National Book Award for nonfiction for *The Hemingses of Monticello: An American Family*.

## WELCOME

I happen to know exactly where I was when the first moon landing took place. I was in Manhattan for the first time in my life, and it was an intimidating experience; at the time, I was running a bookshop in Houston, a city I was familiar with and attached to professionally and emotionally. I had landed at the Pierre Hotel in New York waiting to be reunited with my then-lover. We hadn't seen one another in over two years, and the gravity of the moon landing was somewhat overshadowed by my anticipation of her arrival. The first thing she said to me when I opened the door to my room was, "Did you see them walk on the moon?!?" I was startled but undeterred, and so I gave her the welcome that I'd imagined I would receive from her.

---

**LARRY MCMURTRY** is the author of more than thirty novels (including *Lonesome Dove*), three memoirs, two collections of essays, and more than forty screenplays. He lives in Archer City, Texas.

# SAVE THE MOON!

*Written by* **RYAN BRADLEY** | **WE ARE GOING BACK** to the moon. And when we go back, things are going to be different. We aren't just popping over for a photo op. We aren't going to be there as tourists, spending a few hours on the surface. We are in it, or on it, for the long haul: as a settlement, a mine [or 14], a hotel [or 20], a launch pad to the rest of space. After decades of dreams, ambitions, and false starts, the moon will soon be colonized.

For years now, I've watched as our moon dreams take material form. In a hangar in the Las Vegas suburbs, I gazed upon inflatable space structures funded by motel-chain millionaire and avowed extraterrestrial believer Robert Bigelow. The structures were gargantuan, multi-storied, yet packed down into a crate the size of a Miata. I've seen tests on the Lockheed Martin-designed Orion craft that NASA is banking on to get us back to the moon. I visited the plant where they design, manufacture and test Orion's very particular, very special separation bolts, which explode in a tiny, precise way so that the hunks of rocket or lunar landing system can, when called upon, fall away.

Our moon future is growing ever more real. And not just in America: China has landed on the far side; India will launch a mission to its south pole; Israel is building a new lunar lander; even Nigeria has plans to send out a moon probe by 2030. It all makes sense. Not only is the moon right there in the sky, ripe for the taking, it's now our gateway to the rest of the solar system, and the stepping stone to greater conquests, like Mars.

But as I've been reporting, watching and learning—as I've stood in offices and hangars and factories—there is a part of me that despairs. Inevitably, standing in some office or hangar or factory, an upper manager or company head will tell me about the lunar future he or she is helping to build. This is their dream, they'll say, their ultimate aspiration. But as we stand there, talking about the moon—*the moon*—the language they use is the language of business. It's the return on investment they are after. The pursuit of capital.

It's only when I find the folks on the floor, the people whose hands touch the things that will get us there and back, that I hear a sense of wonder. They'll hold an object and say in a whisper that one day this, or something like it, will make its way onto the moon. To them, the moon is not yet a dollar sign. They still remember that traveling to the moon, setting foot on it—leaving prints that will remain for thousands of years—was once a solemn, sacred task.

The moon is a magical place. It's magical because it's always there, hanging in the sky, constantly changing and yet constant, cycling through its shadows in a steady pattern, longer lasting than our memories or legends, our ancestors and us. The temples we've erected to it, the megaliths, the stones of Stonehenge—none will last as long as the moon. A rock plucked from the lunar surface by David Scott and James Irwin on Apollo 15 is 4.5 billion years old, nearly as old as our solar system. Nothing on Earth is as ancient and unchanging.

Until July 1969, the moon was the place of gods, not men. It was home to the spirits that controlled birth and death and time, change and cycles and seasons, tides and weather and war, mothers and madness and so much else. The Apollo 11 crew knew it was entering the realm of myth and legend. On the fifth day of their mission, as the three-man crew was completing its tenth revolution around the Moon, the craft fell back into transmission range, and mission control came on to deliver the news from back home. Buzz Aldrin, Michael Collins and Neil Armstrong listened to an update that was both inspiring and prosaic: Churches were praying; Genesis was being read from the White House; it was Colombia's independence day; Miss Universe had just been crowned.

"Among the large headlines concerning Apollo this morning," said Ronald Evans, manning the CAPCOM in Houston, "is one asking that you watch for a lovely girl with a big rabbit. An ancient legend says a beautiful Chinese girl called Chang'e has been living there for four thousand years. It seems she was banished to the moon because she stole the pill of immortality from her husband. You might also look for her companion, a large Chinese rabbit, who is easy to spot since he is always standing on his hind feet in the shade of a cinnamon tree. The name of the rabbit is not reported."

"Okay," replied Collins. "We'll keep a close eye out for the bunny girl."

Collins was kidding, but only sort of. [Initial reports ascribed these words to Aldrin, but official transcripts later corrected the attribution.] What had once been the stuff of imagination was now all too tangible. Indeed, by 1969 so much of the astronaut narrative had taken on the contours of a spiritual journey: periods of intensely physical work, followed by an epic quest in the wilderness of space, followed by periods of quiet contemplation in a NASA quarantine. Plus, the confrontation of death, and, upon reentry, a kind of rebirth. "In spaceflight, the experience of one's fears at lift-off, followed by the transition into a wholly different world in orbit, mirrors the death/rebirth cycle," wrote Frank White in *The Overview Effect*, his book cataloging the astronaut experience. "Going into space is certainly a modern metaphor for the journey to higher awareness."

A problem, once earthbound, was putting this higher awareness into words, and answering the eternal question: What did it feel like? There is no good, satisfying answer to this question because words fall short, and astronauts were primarily fighter pilots, not poets. The most fitting response wasn't in words at all, but a painting, by Apollo 12's Al Bean. It's an impressionistic, slightly iridescent sketch he titled *That's How It Felt to Walk on the Moon*.

The other problem with putting a dreamlike place into plain language is that it becomes plain. Armstrong, immediately after delivering his "small step for man" line, began describing the lunar landscape: "Yes, the surface is fine and powdery. I can kick it up loosely with my toe. It does adhere in fine layers, like powdered charcoal, to the sole and sides of my boots. I only go in a small fraction of an inch, maybe an eighth of an inch, but I can see the footprints of my boots and the treads in the fine, sandy particles."

It's these pedestrian bits that now consume our focus. That fine powder, the regolith, is but a layer covering the cold hard assets of raw material that will serve our purposes on Earth and parts beyond.

A few months before the fiftieth anniversary of the first landing, NASA announced that it would send astronauts to the moon's south pole. "We know the South Pole region contains ice and may be rich in other resources based on our observations from orbit," said Steven Clarke, of the Science Mission Directorate at NASA.

The moon, in other words, is no longer the realm of the sacred but a place that is resource-rich. A place we go to for helium 3 and low-gravity kicks. A place for new businesses to launch and new fortunes to be made. This changes what the moon means, our relationship to it.

**TRAVELING TO THE MOON, SETTING FOOT ON IT, LEAVING PRINTS THAT WILL REMAIN FOR THOUSANDS OF YEARS, WAS ONCE A SOLEMN, SACRED TASK.**

Once we've overturned every rock and carved up the whole lunar surface, will it still be the home of gods and elixirs and magical bunnies? Will we still look up at it with fondness? See it as a friend? When the moon is less magical, maybe not magic at all, it's no longer the moon. Just: a moon. A place that becomes, suddenly, Earth-like, in all the wrong ways.

If you doubt this, consider what is happening to Low Earth Orbit, once a similarly magical place of space exploration. It's now a tourist destination, a dangerous junkyard of satellite detritus, and—most recently—the latest piece of real estate for advertising billboards. StartRocket, a Russian company, is currently working on launching satellite ads that will be viewed in the night sky. PepsiCo is an early client. Ads on the moon don't feel all that far off.

Once the moon is more like Earth, our love for it, our stories about it—my stories—will change. I was born on July 20, which for my entire life, as long as I've been aware of it, has been filled with moon-themed gifts, and marked not just by my own passage on Earth but our collective distance from the first steps we made upon that lunar surface.

Several months ago—nine, approximately—my wife and I began calling our son the Moon Dude. He didn't have a name yet, because we hadn't picked one, because he was still in utero. Moon

Dude seemed appropriate. He was, after all, in something like zero gravity, suspended in amniotic fluid, existing in a strange, near incomprehensible space. He was with us but not quite with us. He was our little Man on the Moon. Our Moon Dude.

A few days ago, I took the Moon Dude outside. It was evening. He had arrived on Earth. He has a name, but we still like to think of him as the Moon Dude—there is still so much magic in his very new life. That night was a full April Moon, a Pink Moon, ushering in spring blooms. It was Passover and Easter Weekend—holy days with Moon connections, too, for the heavy tides that helped part the Red Sea; for the three days of the New Moon, symbolizing the three days between Crucifixion and Resurrection.

I told the Moon Dude a bit about these ancient stories, and some even more ancient than that: about Thoth and Isis, Egyptian moon gods of wisdom, and nature and magic; about Artemis, the Greek goddess of the hunt; and Chandra, the Indian god who rides his chariot, the moon, through the night sky while carrying a lotus flower. I told him about Chang'e, who lived in the sky with her husband, an archer named Hou Yi, until they were banished to Earth to live as mere mortals. Hou Yi went on a quest to find the secret to immortality, to get them back into their sky home. He found it, eventually, in the form of a pill, which he put in a box and made Chang'e promise not to open, and which she opened, and ate, and after eating floated her way up and up into the sky, all the way to the moon. She lives there now, with a giant jade rabbit, yes, but also a woodcutter named Wu Gang, who is constantly cutting down a cinnamon tree that continually grows back. He is stuck there for all eternity, cutting down an ever-regenerating tree.

By this time, the moon had come into view through the trees, the Moon Dude's eyes very wide. We stared up in silence for a few moments, me watching both the moon and the Moon Dude, not sure if he could see the orb in the sky but struck nonetheless by the long arch of humans before me who had done this very thing— telling versions of the same old stories to their children, waiting for their old friend the moon to come into view. To lose these moments, that wonder, would be to lose a part of ourselves that we'll never get back.

The irony of our ambitions is that, in returning to the moon, laying claim to it, making it ours, we'll lose the sense of awe that necessitates myth, that makes these stories sing through the ages. The moon may one day quite soon make some of us very rich, as the rest of us lose something of incalculable value.

Meanwhile, as the moon is still magic, I will continue to tell these stories. They belong to all of us, and they are free to tell.

**RYAN BRADLEY** is a writer based in Los Angeles.

## THE SHORTEST NIGHT

I remember it as if it happened in the spring, maybe because in Toluca it never warmed up. But it happened so late it might have coincided with the end of senior year—the last day I ever went to school in my hometown, 40 miles west of Mexico City.

It was July 11, 1991 at 1:24 p.m. when the moon Pac-Manned the sun and all went dark—not black as night, but gray as dawn.

News reports said animals and flowers grew disoriented and went to sleep. It was the shortest night we ever lived—six minutes and fifty-four seconds. That total eclipse of the sun was the longest of the twentieth century. Only on June 13, 2132 will somebody on Earth, perhaps, be able to experience a longer one.

Four weeks after the eclipse, I moved to the galaxy called Mexico City to go to college. Ten years later, I moved farther away. I haven't lived in my home country, my hometown, ever since.

We spilled into the street, my friends, my girlfriend at the time, and me, as daylight waned. I remember our faces paling in awe. The moon going black. The air turning cool and eerie. And us down on Earth—so hopeful and young and alone.

***

**ANTONIO RUIZ-CAMACHO** is the author of the story collection *Barefoot Dogs* and the forthcoming novel *The Healing Room*. He lives in Austin.

# WAXING

*Written by* **RICK BASS** | **THIS IS NOT AN ECLIPSE STORY.** I grew up in Houston in the 50s, 60s and early 70s, so yeah, the moonshot was a pretty big deal. One of the Apollo astronauts was from nearby LaGrange, where my mother grew up— he was not a moonwalker but one of the first ones to get blasted away from this sweet blue planet, out into the unknown darkness, to the moon and around and then back.

My own orbit from those days is widening, casting further and farther toward what some day I suppose will be the territory of old age—outer darkness, with only the cold pinpricks of stars for company, while far below and closer to the warm center run and laugh and play the living. Look how *small* they are all becoming below, the young, as one's orbit widens, with the distance between one year and the next becoming slighter, like the growth rings of an old tree bunching up out at the perimeter, too close together now to even count.

From those childhood days, I remember the moon rocks at the Houston Museum of Natural Science, where my mother would sometimes drive me. In my memory, the exhibit featured a small handful of these rocks on a plain white surface, nondescript in every way. The geologist I would one day become might describe them as clastic, dry, friable, with a low level of compaction—mafic, brecciated igneous—possibly alkaline? They looked almost chalky, like carbonates, though even as a child I noted the obvious lack of fossils. But weren't those tiny vesicles testament to gases, suggesting the possibility of a once-upon-a-time oxygen component? And speaking of air—how did the astronauts do *that*? How could they carry that much air with them, or that much water?

There's an assumption of constancy in Einstein's elegant $E=mc^2$ that seems to mostly get us in trouble. Maybe Einstein meant constancy in the moment, the split second in which the equation is applied, but the string theory people have been saying for a long time that nothing we see is a constant, nor can we be 100 percent sure that it is real, instead calling all matter, all things, an extreme likelihood of quivering assemblages always in motion, assemblages of smaller things arranged to look and feel like the thing that we—ourselves a swirling arrangement of matter—are seeing, touching, smelling, and hearing.

*No ideas but things*, wrote William Carlos Williams, less than a generation after Einstein, not so much resisting as modifying the equation—and it is this dictum that resounded more with me as a boy of ten at the museum. What I remember most, beyond the moon rocks' dirt color, was my desire to *touch* them, and my still dogged if not slightly diminished belief that because they were different, rare, and hard-gotten, they possessed power. They *had* to be treasure.

Surely they contained such power—radioactive or otherwise—that mere Plexiglas could not contain them. I hovered; stalked round and round the squarish stones. Was it a joke? Were the sapphires, moon diamonds, amethyst, and beryllium in vaults somewhere else while the astronauts pranked us with these chunks of road rubble they'd picked up on their way home from the airport?

I drifted away. The other things—the living—pulled me. The aquariums, brilliant with tropical fish; the huge-eyed caiman sunning beneath its heat lamp. No one had a clue then what global warming was. It was upon us but we could not see it. We could see it but we did not notice it. The baby snapping turtle, black as tar, as elegant underwater as a ballet dancer, his long Stegosaurus tail trailing as his oversized feet waved in slow motion. His red eyes were studded with an asterisk for each pupil, leading one to suppose he perceived a different world than the one I beheld; and which of us was to argue the other's reality?

The rocks were an embarrassment. The geologist I would become wished the astronauts had stayed longer, dug deeper, searched farther. Climbed a mountain. Gone around to the back side. To have not returned until they found something better.

My dad—a geophysicist and a hunter—and my mother, a schoolteacher, saw to it that I had a taste for nature not just in the museums but in the woods. We would go hunting every fall in the Texas Hill Country, at a place we called the deer pasture, a wild, feral land of incredible stargazing. Gillespie County. One night, when my brother, B.J., and I were outside, I decided to do something that had long intrigued me: climb nearby X Mountain, even though it was on someone else's property.

In Texas, the feudal notion of private property, the sanctity of personal territory and ownership, is deeply entrenched, ridiculous as it might seem from a biological perspective. The gridwork of fenceposts and barbed wire lattices the entire state as if dicing it into so many croutons. Every strand holds tufts of hair or fur or feathers from all the passers-through except our own kind.

I had been living in Montana long enough by that point to have become more than comfortable with the concept of the commons.

The Hill Country contains some of the oldest stone that can be found at the Earth's surface: Cambrian sandstone, what was once the floor of the new world, the dawn of all life, simple organisms spinning in the sunlit seas—organisms so tiny they can't be seen in the stone in which they now exist, though sometimes, when I'm building a wall or flagstone driveway with those rocks, and I drop one, releasing a wisp of arid dust, I like to think that what I see and smell in that plume is the ground-up ephemera of life a billion years ago: the first scent of us.

The deer pasture also possesses some of the world's oldest granite, rock that's younger than the Cambrian sandstone it pierced from below with its tongues of flame, the mineral-rich magma surging upward along any fissure it could find, the minerals that have existed in the great furnaces near the Earth's dense center demanding their time in the sun.

Some made it out and cooled rapidly; others made it almost all the way out but not quite—and, resting just beneath the surface, the minerals in that magma gradually rearranged themselves according to their polarities and chemical charges and valences, spinning and rotating, their earth-center miles below and the strange moon-rock in the sky, which was maybe related to them or maybe not: and it was in this slow, just-beneath-the-surface cooling that great beauty was achieved. The crystals began adhering to one another, blossoming into fantastic spires and cathedrals, each assemblage of elements having all the time in the world to form, and grow. These are the crystals—the slow crystals—I wanted to believe also comprised the heart and soul, the inner being, of the moon.

B.J. and I set out toward X Mountain. Because it was on the other side of fences, it seemed far away. In reality, it was ludicrously close. At the first fence, I stood on its lowermost strand of barbed wire and, expecting there to be some give, lifted my other foot to cross. But the wire was newly strung and too taut, and there was no stretch, so I wobbled, raking my calf across one of the fence's twisted teeth, the single barb sharper than a knife. It was the kind of wound that cuts so cleanly there is no pain, only the sudden tickle of blood's wetness on skin. The blood went from warm to cold quickly as it trickled into my sock.

In his great song "Lake Marie," John Prine pauses his singing to query the listener: "You know what blood looks like in a black-and-white video? *Shadows*! That's what it looks like. *Shadows*." I'd never quite understood the allusion, but looking at my leg in that hyper-brilliant silver-blue light, I was reminded of that song. The blood had the gleaming quality of old-time flashbulbs. And it was definitely the darkest thing in that mercuric, floodlit world: the only dark thing, I realized, which might have been what Prine was seeing and describing.

The tear—the slice, slash, gash—was right in the meat of the calf muscle. *That one's gonna leave a scar*, I thought, and as we proceeded on toward the hill—the mythic mesa we had seen all our lives but which the colonists took from Mexico, who took it from the Comanches, who may or may not have taken it from someone before them, the Athabaskans?—it grew smaller. We passed through a grove of oaks, the shadows as dark as the trunks of the trees themselves, and I left a smear of blood on blades of grass and on the silver-fire leaves of agarita and shin oak.

We started up the admittedly steep slope, but with every step, the mountain before us shrank until it could not even really be called a hill. It was a flat-topped bump—a little neck of caliche, limestone, a remnant of older sea-times, compressed to chalk. We stood on it, looked out. I felt like a child. I was, what, maybe 40 years old? Perhaps not even.

The moonlight bathed us among the wind-blasted juniper. I know this is a wretched cliché, moonlight *bathing*, but it's true, it poured down and over us as if molten silver. It was brighter than most daylight, yet there was no danger of moonburn.

The reversal in scale—the grand becoming almost minute—made me dizzy, as did the moonlight itself. Back when photographs were taken with cameras, not phones, and you took the film to a print shop for developing, you had to look at the strip of negatives to decide which reprints you wanted. Dark became ghostly bright, and light became unseeable dark. I felt that I was getting a glimpse of the way the world really was, not to me but

**WE PASSED THROUGH A GROVE OF OAKS, THE SHADOWS AS DARK AS THE TRUNKS OF THE TREES THEMSELVES.**

to someone—someone else's reality—and whether that was raccoon or scorpion, bumblebee or night-blooming cirrus, hummingbird or swan, I could not say, only that we were in it. The top of this great mountain—visible from 30 miles away—was not much larger than a suburbanite's lawn. It was the size of a burger joint's gravel parking lot. And yet: it was so level, in a land where nothing else was. We walked around, feeling much closer to the moon now—hundreds of miles closer, rather than a hundred feet. A jackrabbit, pale as bone, looking like a snowshoe hare in winter, leapt from hiding and dashed away. Then, at the southern end of the mesa, I noticed something: smooth white round riverstones, spaced evenly in an arc.

They were grown over with grass and lichens, but the moonlight brought them out in bony relief. Now I could see more, each no larger than a skull, but enough of them, I realized, to form a circle. The circle was grown over with low juniper, but it was, by god, a real teepee ring—which made sense to me, though I'd never seen one in these parts. What was 100, or 125 years, to a stone, or even the placement of a stone?

I didn't mind being up there, uninvited by the absentee landowner, who simply had his cattle grazing below, but I was a little rattled by having stumbled into a ceremonial spot uninvited. It had long been unused, of course, but still: the incredible light made me feel super-illuminated in a way I did not want to be, and I apologized for barging in, and we made our way back down toward the fence, with my leg still painting the low vegetation bright red.

We had not gone far at all when we encountered an immense white-tailed buck, his velveted antlers glowing, as if he carried above him a silver nest of fire. Big bucks like that are always nocturnal, but this one looked uncomfortable—as if he was being called upon to *swim* through that silver light, which was so strange and thick it felt like a chunk of matter, elemental like a mineral rather than waves—and, after watching us for a few seconds, turned and ran and vaulted high over the fence, arching like a rainbow. He landed lightly and continued on into the juniper, bobbing like a spark: the living, taking refuge in the living.

We approached the same illuminated fence, touched it first, as if it might be hot, and then climbed carefully over. The Old Ones who had sat on that hill—who were they, how long had they sat there, what thoughts had they considered? Where were they now, and will each of us one day become just as invisible? It seemed to me that if one lived, burned, intensely enough one might exist as a kind of echo, or shadow: in the way that, in beholding the mirror of the moon, we are seeing an echo of the sunlight that was cast many years ago—light-years—and which, reflected, falls down upon us, the echo of an echo, encasing us as if in amber.

More than a year before the solar eclipse crossed the American landscape in the summer of 2018—not from east to west, in the style of Manifest Destiny, but reversing the curse, many hoped, from west to east—the lucky farmers whose pastures fell within its path erected a frenzy of billboards to advertise parking spots and viewing locations. Such was their anticipation that the hand-painted plywood became sun-faded long before the event, as if the eclipse had already come and gone, or as if the eclipse itself had aged the signs, their red letters—*Parking*, $5.00—blurry now, wavering like old bloodstains.

As the months melted and we were all pulled closer to the day of reckoning, a sweet kind of unification seemed to happen: the mass of us becoming aware of the time, date, location—the countdown— our minds adjusting like crystals in cooling magma, aligning and then converging from all directions to behold the approaching singularity.

I considered taking off work to hie down toward Bitterroot or Gallatin country—about an eight-hour drive—rather than stay in my home valley, where the show would be only partial. But I wanted to feel what my valley felt in that darkened hour. I wanted to note any shift in the wind, any skip or stutter in the pull of gravity; if the calls of ravens changed, whether the thrushes—crepuscular singers, lovers of the gloom and gloaming—began to sing. I also—like almost everyone, I think—did not want to experience it alone.

For weeks, community service organizations—libraries, notably, and other do-gooders—had been passing out free eclipse sunglasses, but having procrastinated until the day before, I called around only to discover all supplies had run out. The next-best-thing, I imagined, might be a welder's helmet, so I drove to the hardware store in town an hour before closing. The one helmet left was too expensive, so, inspired by necessity, I found some replacement glass sheets for welder's masks. Each sheet was dark green, smaller than an index card; I bought three.

The day of, I drove downvalley to the office of the conservation nonprofit I work with. Because I'd read that it was safer to watch the eclipse reflected in a body of water, rather than staring at it directly, I found a child's blue plastic swimming pool and filled it with water. The staff and I then went out into the backyard. We stared at the pool as if awaiting the emergence of the Loch Ness monster; glanced sidelong now and again, up at the same old sun up in the same old sky. Same old birdsong. The in-between time, in the north country: summer winding down, autumn not yet arrived. Torpor.

It came slowly. There was a blurring, a wavering, that was, to be honest, a little unsettling: more so, I think, than the coming shadow.

To have seen a thing one way all one's life, for 60 years, then to see it, the previously immutable, waver and sprawl—well, what if everything contained such waver, such wobble? As if our very existence—once loosened—could unravel, back into the nothing from which we arose.

The edges shimmered in the way that waves of heat rise from pavement in deep summer. *Holy shit*, I remember thinking. It didn't get dark so much as fuzzy. There appeared to be a kind of static in the air—the visual equivalent of the itchiness or scratchiness of a wool jacket; as if a coarser weave of pixels were registering on our brain. We glanced again and again up at the sun, then into the pool, where the image was indeed more distinct.

**THE SAVVY ONES DID NOT STAY AT HOME TO WATCH THE ECLIPSE LIKE ME, BUT WENT DEEP INTO THE WILDERNESS TO BE BATHED IN FULL DARKNESS**

Once, I saw the black silhouette of a witch on a broom riding the crescent black moon across the face of the sun—but when I looked in the pool, the witch disappeared.

I heard a car approaching. I saw it was the mail lady, in her white jeep, the flashing wide-load strobe light atop. Amazingly, she had not heard about the eclipse, so I asked if she wanted to join us. She turned the jeep off. I could imagine that driving the same route decade after decade was a little numbing—always driving, never walking, 150 miles roundtrip, six days a week. I handed her the magic panes of glass, told her not to look up without them. Told her to look into the pool first.

The day was not exceptionally dark. Instead, its dominant characteristic was stillness—the stillness of hesitation. As if not only were the humans entranced but also everything with a heart or a spirit within, an essence that pulsed and throbbed. The mail lady held the strips of glass carefully—daintily—her smoker's stained fingertips suddenly elegant. She stared into the pool as, above us, earth, moon, and sun continued to separate.

We were all being returned to our old ways. In the forest, the birds were making their little late-morning sounds, but—and I acknowledge this may be only my interpretation—they seemed to be a little tentative, indecisive. I stood beneath the static, the strange dim light that wasn't darkness. I felt cleansed, somehow, lighter. Not so much forgiven as—cleaner. Child-like.

The mail lady stared into the pool so long that I started to wonder if she'd had a spell cast on her; that she might have decided to never again deliver the mail. That the little glass plates, unused, would fall from her hands; that she would sit, like the woman staring up the hill in Andrew Wyeth's *Christina's World*, and—content now—lead a life of such monastic dedication to the baby pool as to forego food and even the pool's sun-black water. I wondered if we had saved her vision: if, driving upriver, listening to music, she might have stared at the black witch on the black broom riding the black moon across the sun and had her eyesight so damaged that she could no longer drive; if she might have been unable to sort letters any more; if hundreds and thousands of pieces of mail might have been misdelivered with devastating consequences—I inquired meekly what she thought of it all. She looked up, surprised to see me. The witch was on the back side of the sun now, somewhere invisible in all that blue sky. Birdsong did not exactly erupt, but I felt—we all felt—the world's gears begin to move again. The staff began drifting back to the office, back to their work of saving the forests, the mountains, that are, for now, our home.

"That was something," the mail lady said, looking back into the water. She, too, was returning to her old self, but more slowly than the rest of us, as if she had been taken back, way back, to some point earlier in her life: young adulthood, or even more distant—back to a time when, setting out each day, one not only expected to see such trippy phenomena but also sought them out. She walked—shakily, I thought—back to her jeep, turned on the flashing amber lights, and continued on up the road: changed, lightened, leavened, undone, remade, like every thing, and every one of us, 99.9 percent certain about almost anything; and diminished, I think, for that excessive belief, that confidence and security. Blinded, even, maybe.

I sat by the side of the road like a wayfarer, not so much waiting as decompressing from what I had seen, and the distance I had traveled. The earth to which I had returned. I felt extraordinarily calm. I felt ready to start again. Felt my old self, isolated, but wanting to assemble—to find beauty, as Terry Tempest Williams says, in a broken world.

And to disassemble: wanting to stretch, if not fully unravel, the vertical, humming strings of matter—like beaded curtains—that physicists tell us represent the percentages and probabilities of reality; to test the almost-certain quality of it, and in so doing, maybe sometimes get a peek at what might be beyond that veil.

I heard a car approaching: other than the mail lady's jeep, the first one all morning on this strange and bestilled day. It occurred to me that, other than our staff and the postmistress, I'd not seen another human, and that it had likely been the same for this driver.

This world is beautiful but it is never quite finished. One can always push against, sand or polish—or prune or shave—its furthest edges. Without even really knowing what I was doing, I stood up, stretched both arms out in a zombie pose, and began walking away from the road, Frankenstein-like, toward the woods, trapped by the light, as the car and driver zoomed past and, hopefully, if for even just a moment, wondered at what he or she was seeing, before being drawn farther up the road by the shimmering curtains that keep us all moving forward.

The savvy ones did not stay at home to watch the eclipse like me, but went deep into the wilderness to be bathed in full darkness, only the edges of all things limned with a corona of fire. Do they know the answer now, if only subconsciously, while I still search? When we see, are we really seeing? We know that whatever we are looking at is 99.9 percent "real," or true—but what does 100 percent look like? Does it exist in the moon's shadow? Or does this strange planet of us, in those 90 or so minutes of a total eclipse, begin to unravel—still real, still true, but with stuttering images of a further reality?

Much of what we behold—that which we have made and woven— is as but a dream, surreal and even unreal. As if where we began—on the platform of the old stone, at the edge of an old sea, and craving light, craving shelter and protection, craving food from the garden, craving craving craving—was the real thing. As if the old implacable stone is the truth and the light; the quivering, shimmering likelihood of us, as we exist or mostly exist right now, an extreme probability. We move around in the light, but we cannot yet see. But how we crave, more than ever, those five points of attachment: touch, taste, scent, sight, sound.

Come back, rock. Come back, moon.

---

**RICK BASS** is the author of more than thirty books of fiction and nonfiction. He lives in northwest Montana, where he is a board member of the Yaak Valley Forest Council and Save the Yellowstone Grizzly and teaches writing workshops.

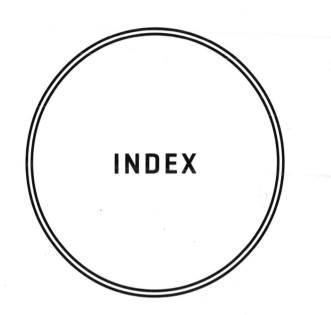

INDEX

# INDEX